The Art of the "What-If?"

AF705502

Paul J. Nahin

The Art of the "What-If?"

In 17 Science Fiction Tales from the Past

Springer

Paul J. Nahin
University of New Hampshire
Exeter, NH, USA

ISBN 978-3-032-15617-4 ISBN 978-3-032-15618-1 (eBook)
https://doi.org/10.1007/978-3-032-15618-1

© The Editor(s) (if applicable) and The Author(s), under exclusive license to Springer Nature Switzerland AG 2026

This work is subject to copyright. All rights are solely and exclusively licensed by the Publisher, whether the whole or part of the material is concerned, specifically the rights of translation, reprinting, reuse of illustrations, recitation, broadcasting, reproduction on microfilms or in any other physical way, and transmission or information storage and retrieval, electronic adaptation, computer software, or by similar or dissimilar methodology now known or hereafter developed.
The use of general descriptive names, registered names, trademarks, service marks, etc. in this publication does not imply, even in the absence of a specific statement, that such names are exempt from the relevant protective laws and regulations and therefore free for general use.
The publisher, the authors and the editors are safe to assume that the advice and information in this book are believed to be true and accurate at the date of publication. Neither the publisher nor the authors or the editors give a warranty, expressed or implied, with respect to the material contained herein or for any errors or omissions that may have been made. The publisher remains neutral with regard to jurisdictional claims in published maps and institutional affiliations.

This Springer imprint is published by the registered company Springer Nature Switzerland AG
The registered company address is: Gewerbestrasse 11, 6330 Cham, Switzerland

If disposing of this product, please recycle the paper.

Dating from 1844, "The Infinity Juggler" illustrates the idea of an infinity of possible worlds, which would become (and remains) a popular theme in science fiction. Painting by "J. J. Grandville," the professional name of the French artist Jean-Ignace-Isidore Gérard (1803–1847).

Also By Paul J. Nahin

Oliver Heaviside (1988, 2002)
Time Machines (1993, 1999)
The Science of Radio (1996, 2001)
Time Travel (1997, 2011)
An Imaginary Tale (1998, 2007, 2010)
Duelling Idiots (2000, 2002)
When Least Is Best (2004, 2007, 2021)
Dr. Euler's Fabulous Formula (2006, 2011)
Chases and Escapes (2007, 2012)
Digital Dice (2008, 2013)
Mrs. Perkins's Electric Quilt (2009)
Number-Crunching (2011)
The Logician and the Engineer (2012, 2017)
Will You Be Alive Ten Years From Now? (2013)
Holy Sci-Fi! (2014)
Inside Interesting Integrals (2015, 2020)
In Praise of Simple Physics (2016, 2017)
Time Machine Tales (2017)
How to Fall Slower Than Gravity (2018)
Transients for Electrical Engineers (2019)
Hot Molecules, Cold Electrons (2020, 2022)
In Pursuit of Zeta-3 (2021, 2023)
The Probability Integral (2023)
The Mathematical Radio (2024)
The Delicate Art of Brute Force (2026)

Prelude: For Those with Math Anxiety

Even when he didn't follow what he was looking at, he could relish a good page of mathematics all day long.

—The admirable attitude of a character (who, we are told, despised what often passed for literature) in John le Carré's 1989 spy novel *The Russia House*.

Calculus is an art—it endows you with wonderful powers; you can let your imagination go to all sorts of lengths and not pass out of the realm of reality. Calculus is like chocolate meringues—elementary trig is [by comparison] like very thick stale bread with margarine.

—From a letter dated September 12, 1923, by Barnes Wallis[1] to his future wife.

In this book, I discuss a number of science fiction and fantasy tales that explore various "what-if?" questions about possible futures. They are tales in which, along with actual magic, or with situations that may *seem* like magic, there also appear *mathematical and physical* (and even *philosophical*) ideas. Ideas that are, in fact, central to the story (in some cases in ways that even the authors themselves might have missed). With few exceptions, there are no actual, detailed mathematical derivations here, and I mostly limit the little mathematics that is here to simply quoting results and (usually) providing

[1] The British aeronautical engineer Barnes Wallis (1887–1979) was the inventor of the back-spin bouncing bomb that the RAF used in May 1943 to breach the Möhne and Eder dams in the German Ruhr Valley. That seemingly science fiction/fantasy, but true, Second World War episode, was made into the 1955 movie *The Dam Busters*, and a bouncing bomb attack on a dam makes a dramatic appearance in the 2025 video game *Sniper Elite: Resistance*. The quotation is from the 2005 book *Mathematics with Love* by Wallis' daughter, Mary Stopes-Roe.

citations to other works where, if you are interested, you *can* find derivations.

One feature of this book, that will definitely tell you that it isn't a typical book *about* science fiction and fantasy, in the usual sense, is that at the end of nearly every chapter is a fictional tale to illustrate the central idea of that chapter. That is, included here are *actual stories*, not just descriptions of stories. Using fiction in a book dealing with technical matters *is* unusual, yes, but not unheard of: the great success of Clifton Fadiman's *Fantasia Mathematica* (Simon and Schuster 1958) and its sequel *The Mathematical Magpie* (Simon and Schuster 1962) are examples of that, as is H. Bruce Franklin's scholarly *Future Perfect* (Oxford 1978).[2] This book follows Fadiman's and Franklin's approach of using fiction, but with one not-so-minor twist. They used the fiction of numerous authors, while in this book *all* the fiction is from the same author—me.

I did this for one very practical reason. To use the fiction of any author, you must be quite careful not to violate copyright protections, which involves getting written permission from the copyright holder (something not always easy to determine, especially for deceased authors), along with (almost always) the payment of a reprint fee. Violating copyright is to invite all sorts of nasty legal action, along with possibly large financial penalties. *Blatantly egregious* violations could, if you *really* get on the wrong side of a copyright holder, result in jail time! The bottom line is that the intellectual property courts take the theft of somebody else's creativity very seriously.

Using my own stories, however—all of which were sold to and published in various science fiction/fantasy magazines, or to the editors of book anthologies—nicely avoids all such unpleasant complications. In every case, I sold *only* the first-publication rights. After its publication all rights to a story reverted to me, rights which I have (with no charge!) offered to myself.

The use of stories is not some sort of sneaky way to imply that hard-core math and physics, themselves, are "boring" (or "stale," to use Wallis' word) and have no magical nature to them.[3] They certainly do! So, as the first of

[2] More recently, one of the best modern treatments of the interaction of science with science fiction is John Cheng, *Astounding Wonder: Imagining Science and Science Fiction in Interwar America*, University of Pennsylvania Press 2012. Cheng's book is an excellent study of the cultural relationship between science and science fiction.

[3] The word "magic" doesn't necessarily demand the violation of physical law. The science fiction writer Arthur C. Clarke (1917–2008) is famous for (among many things) the following assertion: "Any sufficiently advanced technology will *seem* [my emphasis] to be magic." As an example of that, try to imagine what listening, for the very first time, to a radio in 1922 (the start of broadcast radio) must have been like for somebody born at the start of the twentieth century. For such people, even as teenagers, there was no radio. Then, *suddenly*, there were the voices of famous people speaking and singing coming out of talking

those "few exceptions" I mentioned a moment ago, let me elaborate just a bit on what I mean by that.

I was once invited by a math teacher at my local high school to appear in his algebra class as a guest speaker, where I could talk about anything I wanted (as long as it was about math!). What follows is what I did during my visit, starting with an easy question. The math you'll see here is what I talked about in that guest appearance, and the way it is presented here will give you a good idea of the way the little math that is in the rest of this book is handled. For those who are not mathematicians at heart, I think you will be reassured that this is *not* a mathematical treatise.

I began my talk by asking the students to consider the following situation, our first "what-if?" question in this book. Imagine a man who goes on a curious walk. One that sounds like something you'd find in a science fiction story. He starts by taking one step to the right of one foot. Then, a second step further to the right of one-half a foot. Then, a third step to the right of one-fourth of a foot. And so on, forever, with each new step being one-half of the previous step. How far from his starting point does he end-up? This was, I realized, almost surely a puzzle that most high school math students had seen before, but I intentionally started with it to "set-up" the class for what I suspected *would* be a jaw-dropping surprise for them.

If, as a mathematician would, we let S_1 represent the sum of all those individual steps, we would write (to the joy of that character in *The Russia House*):

$$S_1 = 1 + \frac{1}{2} + \frac{1}{4} + \frac{1}{8} + \ldots$$

where the three dots mean "and so on, forever." What we do next is some simple first-year high school algebra (I could see that most of the class was already ahead of me with that) to first factor $\frac{1}{2}$ out of the summation on the right to get

$$S_1 = \frac{1}{2}\left(2 + 1 + \frac{1}{2} + \frac{1}{4} + \ldots\right) = \frac{1}{2}(2 + S_1) = 1 + \frac{1}{2}S_1.$$

Next, subtracting $\frac{1}{2}S_1$ from each side of the equation we arrive, just like that, at

boxes in the kitchens, bedrooms, and living rooms of homes all over the world. Magic indeed! For much elaboration on "radio magic," see my *The Mathematical Radio*, Princeton 2024.

$$\frac{1}{2}S_1 = 1$$

and so

$$S_1 = 2.$$

The sum of an *infinity* of his ever-smaller steps has taken the man just 2 feet from where he started. The students nodded their heads in agreement (just as I had hoped) and I could sense that the general feeling was one of "this is so old-hat." Well, I thought, I will soon disabuse them of *that* notion, as I was now ready to spring my surprise! A surprise that forms our next "what-if?" scenario. What now, I asked the students, if the man alters his walk to take a different yet still infinite in extent sequence of ever-decreasing steps, a walk described by the sum

$$S_2 = 1 + \frac{1}{2} + \frac{1}{3} + \frac{1}{4} + \frac{1}{5} + \frac{1}{6} + \frac{1}{7} + \frac{1}{8} + \frac{1}{9} + \ldots = ?$$

Where does the man now end-up? Obviously, I asserted, the value of S_2 will be larger than that of S_1 because S_2 has all the terms of S_1 *plus* a whole lot more. When I asked the class to guess the value of S_2, I got answers starting from 3, but never more than 4. One particularly bold student suggested pi (=3.14159265…). "Well," I told them, "*None* of those guesses are right. What would you say if I told you the answer is, in fact, *infinity*?"

That got (as I hoped!) snorts of disbelief from most of the students, with the bold one exclaiming "No way, man! All the steps are, as before, getting shorter and shorter, so how do you get infinity out of that?" In reply, I showed him and his fellow students a wonderfully elegant argument due to the fourteenth century French religious scholar Nicole Oresme (1325–1382). He started by simply writing the expression for S_2 with an explicit grouping of the terms:

$$S_2 = 1 + \left(\frac{1}{2}\right) + \left(\frac{1}{3} + \frac{1}{4}\right) + \left(\frac{1}{5} + \frac{1}{6} + \frac{1}{7} + \frac{1}{8}\right)$$
$$+ \left(\frac{1}{9} + \frac{1}{10} + \frac{1}{11} + \frac{1}{12} + \frac{1}{13} + \frac{1}{14} + \frac{1}{15} + \frac{1}{16}\right) + \ldots$$

where each grouping has twice as many terms as the previous grouping. We can clearly do that without end, as the integers go on forever. Then, in each grouping, replace every term in that grouping with the last (the *smallest*) term in that grouping, to get a total sum that is clearly *smaller* than S_2. That is, Oresme wrote

$$S_2 > 1 + \left(\frac{1}{2}\right) + \left(\frac{1}{4} + \frac{1}{4}\right) + \left(\frac{1}{8} + \frac{1}{8} + \frac{1}{8} + \frac{1}{8}\right) + \left(\frac{1}{16} + \frac{1}{16} + \frac{1}{16} + \frac{1}{16} + \frac{1}{16} + \frac{1}{16} + \frac{1}{16} + \frac{1}{16}\right) + \ldots$$

or

$$S_2 > 1 + \left(\frac{1}{2}\right) + \left(\frac{1}{2}\right) + \left(\frac{1}{2}\right) + \left(\frac{1}{2}\right) + \ldots.$$

In other words, S_2 is bigger than 1 plus an *infinity* of $\frac{1}{2}'$s, and so $S_2 = \infty$. The man, even though he still takes ever-smaller individual steps, has now gone on a *very* long walk! Isn't that beautiful? This is what I mean by saying "math is magic" because it lets us calculate the answer to such a curious question. Oresme's result is, in fact, not the end of the magic. To see that, suppose the man modifies his walk once more so that, after the first step, he turns to his left through 90° and *then* takes his second step of one-half of a foot. And then he turns another 90° to his left before he takes his third step of one-third of a foot. And so on, always turning 90° to his left before taking the next step. How far from his starting point does he end-up? As it turns out, not very far, at all. To prove this takes a bit more math than does Oresme's proof but, actually, no more than what a high school senior who has taken AP-calculus would have (don't worry about this—we aren't going to do *that* here!).

However, to simply *see* that the man doesn't end-up very far from his starting point is actually easy, *if* you sketch a diagram of his walk, as in Fig. 1. There, it is physically clear that he *spirals in* toward some point in the first quadrant of the x, y two-dimensional plane. Figure 1 makes it obvious that the man's walk can be described with *two* sums: S_x (the sum of his steps in the back-and-forth x-direction) and S_y (the sum of his steps in the up-and-down y-direction). That is,

$$S_x = 1 - \frac{1}{3} + \frac{1}{5} - \frac{1}{7} + \ldots$$

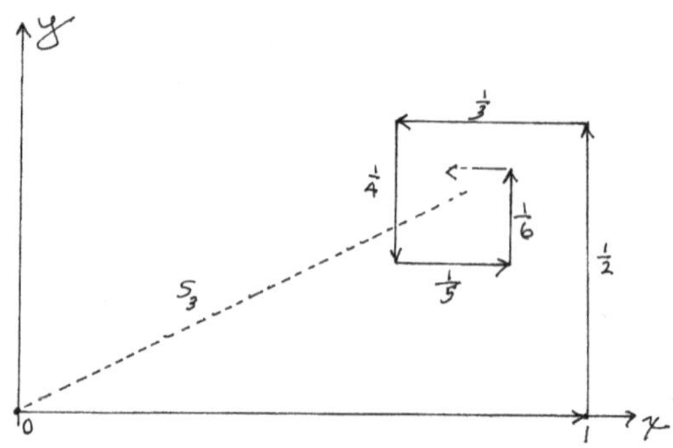

Fig. 1 The man's third "what-if?" walk of an infinity of ever-smaller steps

and

$$S_y = \frac{1}{2} - \frac{1}{4} + \frac{1}{6} - \frac{1}{8} + \ldots = \frac{1}{2}\left(1 - \frac{1}{2} + \frac{1}{3} - \frac{1}{4} + \ldots\right)$$

and thus the distance from his starting point to where he ends-up is given by the Pythagorean theorem to be

$$S_3 = \sqrt{S_x^2 + S_y^2}.$$

The math questions I then posed to the class were the obvious ones: what are the values of S_x and S_y? The answers to those questions illustrate the point where I will draw the line in this book at foregoing any detailed math derivations. What I did for the class is what I will do here for you—simply tell you the answers and indicate where you can find a derivation *if you are interested*. So,

$$S_x = \frac{\pi}{4} = 0.7853\ldots$$

and

$$S_y = \frac{1}{2}\ln(2) = 0.3465\ldots.$$

Thus, from the Pythagorean theorem, we have $S_3 = 0.8584$ feet and we see that the man ends-up less than one foot from his starting point.[4] These amazing results illustrate the truth in the words of a mobile infantry combat soldier when musing on the value of mathematics (in the 1957 novel *Starship Troopers* by Robert Heinlein (1907–1988)): "[I]t doesn't hurt to learn all you can of it … everything of any importance is founded on mathematics."

The discussions in this book are, for the most part, quite serious, and I will, *usually*, treat physics and math with all due respect. But that doesn't mean we have to be relentlessly humorless. One unusual feature of this book is, as I said earlier, that nearly every chapter ends with a *fictional* treatment of the central topic of that chapter. As the first illustration of that feature, what follows is a short tale that can only be described as a *pure fantasy*, written simply for fun, in the same spirit of the calculations concerning imaginary walks with an infinite number of steps. It originally appeared in the August 1981 issue of *Rod Serling's The Twilight Zone Magazine* and, while it has no math,[5] it is almost certainly the most outrageous example of imaginative fiction in this book (after reading it, I believe you will agree), and I think it a perfect way to convince you that this is *not* a typical technical book!

"The Next Time Around"

Rollo Adams pulled out of the motel parking lot just before dawn. It was best to hit the smooth, hard pavement of the superhighway while the air was still cold and the concrete slightly wet with dew. The souped-up convertible accelerated quickly and smoothly to seventy, and Adams settled in for last, long day of cross-country travel.

As he watched the flat emptiness of the Arizona desert flash by, he felt the wind blow over his tiny bald spot, The old carcass sure did ache! His left side still hurt where the vandal had kicked him yesterday. If somebody hadn't come along just then, the bastard probably would have stuck him with a

---

[4] As pure math problems (that is, ones with no physical interpretations as "what-if?" walks consisting of an infinity of steps), the values of S_x and S_y have been known for centuries. S_x was discovered in 1671 by the Scottish mathematician James Gregory (1638–1675), and S_y was found in 1668 by the Dane Nicholas Mercator (1628–1687). If you are curious about how to do such calculations, write to me at paul.nahin@unh.edu and I will send you derivations using just high school AP-calculus.

[5] It was at least partially *motivated* by math, however. One late night, nearly a half-century ago, I was sitting in my office grading exam papers from my engineering probability class. After getting about half-way through a maze of scrawled analyses from my students, and having drunk far too much really awful machine-dispensed coffee, I decided I needed a break to recover my wits. What better way to do that than to take 30 min to write "The Next Time Around"?

knife. Damn him! He tried to forget the discomfort by thinking of his destination—the romantic waterfront of San Diego Bay.

The car backfired once, and he wished he had a tachometer so he'd know how many RPMs he was pulling.

The miles pounded by, and Adams—who, as usual in the morning, had felt worn-out, down, deflated—began to feel a good deal better. The heat of the drive smoothed his stiffness away, and his grip on things was becoming firm again. Thank God the episodes of nausea and dizziness were getting less frequent and severe! For a while he'd been wandering, slipping a bit, but he was gradually gaining experience, and he could handle the pressures of his new life better now.

And what pressures! At first, he'd been almost paranoiac about it, always on the watch for danger, never knowing when he might get drilled full of holes. He'd watched all the usual cops and robbers shows on television, sure, but that was just recycled Hollywood fantasy. Now he *really* knew what it meant to be on the run for the rest of his life.

But the thick steel belts he wore around his vital areas reassured him. With body armor like that, it'd take a mighty big slug to rip him open! He had a lot of miles of experience on him now. He'd survived some pretty rough banging around these last few weeks and had learned how well he could bounce back. He'd been pleasantly surprised. He knew he wouldn't blow and lose control. He was tougher, more resilient, than he'd thought.

The road ahead was empty, a ribbon running long and straight to the horizon, and so he let his thoughts drift back to last month, when he'd experienced the most traumatic event in his life. Man, the only creature on earth to be aware of the inevitability of his own death, still learns to cope with it. But it's one thing to read of the passing of a stranger, or even of one casually known; it's another when death strikes closer to home and snatches away your wife.

A giant wave of loss swept over Adams as he thought of Sally. God, how he missed her! He sighed quietly to himself as he realized how the two of them had often joked about what came after death. Crazy things, like coming back as some other person. Sally had always said she wanted to return as a lizard and bake all day on a rock in the sun; she had never liked the cold winters in New England. Adams had chuckled at the thought of his elegant wife sitting on a rock eating flies. When he'd mentioned that to her, she'd frowned momentarily and then declared that it didn't matter, because once she was a lizard, she'd *like* flies!

Half hypnotized by the rhythmic undulation of the road surface, and with nothing else to distract him, Adams continued to remember. He recalled how

Sally had laughed at him when he'd suggested that she might come back all right, but not as a lizard. Maybe she'd come back as the rock! That would be okay, too, she'd replied, tears of silliness running down her face, just as long as she could roast in the sun!

Rollo Adams prayed with all his will that his wife had been granted her request. She'd been so young and beautiful to die in that plane crash; he hoped that her wish had come true. And maybe it had! After all, weren't there religions that said you came back as a higher form if you'd led a virtuous life, and a lower one if you hadn't been so good? Who was to say if a lizard, or even a rock, was higher or lower than man?

Perhaps things had worked out for her; perhaps not. They certainly hadn't for him. Maybe those three or four one-night flings a few years ago were the cause of *his* fate. He'd worshiped Sally, and those few moments of weakness still shamed him. He felt an urge to weep and almost came undone right there. But then his new strength saved him, he had learned, over the past few weeks, to hold everything inside. To let it all out now would be disastrous.

Rollo Adams, dead in the same crash as Sally, roared down the highway. Instantly responsive to the rear-axle, high-torque differential shaft that spun him, he gripped his sporty rear magnesium rim, dug his zigzag slip-proof treads into the road, and felt the pavement rush past beneath him. The road stretched ahead in the hot sun, and San Diego beckoned.

[After reading this, there should be no doubt in your mind about recognizing a *fantasy*!]

This story may at first strike you as having "gone too far" but, in fact, its idea of "coming back" is found in earlier (dating from the 1960s) stories by Philip José Farmer (1918–2009). Those stories form Farmer's famous *Riverworld* series, in which we read of how billions of humans, from all across the ages, have mysteriously been resurrected along the shores of a river millions of miles long on an equally mysterious planet. Now *that* is a premise that goes *far* beyond "The Next Time Around"! Farmer takes the "what-if?" question about resurrection and develops the issues inherent in it to an astonishing level.

Now, two final comments to end this Prelude. First, there is one sort of "math" story that is *not* within the scope of this book: stories dealing with the *psychology* of doing math, of the intimate interaction of the details of a difficult problem with the *personality* of a mathematician attempting to solve that problem. Such tales, when from the pens of insightful authors, can provide powerful reading experiences, but they have no connection with either science fiction or fantasy. A novel-length example of what I have in mind is the 2016 *A Doubter's Almanac* by Ethan Canin, the story of Milo, a professor in the math department at Princeton. Milo is a genius, as well as a drunk, a bad combination the novel explores in depth. Great reading, yes, but it isn't science fiction or fantasy, but rather hard-core reality.

Other examples of such tales are several beautiful short stories by Leonard Michaels (late professor of English at the University of California/Berkeley), about the mathematician Raphael Nachman. Those tales, written between 1997 and 2003 (the year of Michaels' death), have almost no direct connection with mathematics (with one exception),[6] but rather explore the special quirks of Nachman's personality, quirks that play crucial roles in his behavior when faced with situations that occur in everyday life. The Nachman stories were published in *The Threepenny Review*, *Partisan Review*, and *The New Yorker*, journals about as far from the science fiction and fantasy pulp magazines as you can get.[7]

As my second concluding comment, the reason I have talked so much about *mathematics* in this Prelude is because it paves the way for me to introduce a *technical* distinction between science fiction and fantasy. A *science fiction* story allows, at most, *one* violation of a presently accepted law of nature. You can, for example, have a time machine in a story *or* a faster-than-light spaceship but you can't have both.[8] (Unless you're an Isaac Asimov, or a Ray Bradbury, or someone else of equal stature—*then* you can do anything you want!) A *fantasy* story, however, can violate known laws of nature as many times as desired (that's called *magic*!).

In contrast to both science fiction and fantasy tales, *real mathematics* forbids violations of *any* kind, period. And, of course, if a physicist ever does construct a time machine, it is simply inconceivable that his lab notebook would not contain page-after-page of abstract mathematics.

---

[6] The exception is "The Penultimate Conjecture." Nachman has long been fascinated by the Penultimate Conjecture (we never get any details) that nobody has been able to either confirm or refute. Then, suddenly, another mathematician announces he has succeeded. Nachman attends that mathematician's lecture on his proof and, while listening, spots a fatal error. Should Nachman say something (thus humiliating the speaker), or should he keep silent? Nachman struggles with the choice.

[7] The *pulps* got their name from their production on cheap wood pulp paper that, along with the development of high-speed printing press technology, made the publication of "fiction for the masses" economically feasible. An unfortunate result of using pulp paper, however, is that the acid-based paper relentlessly oxidizes into ashes. Today all the pre-1960s pulp stories, the good ones and the not-so-good, have literally burned-up and vanished and exist only in the archives of those libraries that have them on microfilm. As Raymond Chandler (1888–1959), who started his literary career writing for the mystery pulp *The Black Mask*, and who eventually created the hard-boiled private detective Philip Marlowe (e.g., *The Big Sleep*, made into a 1946 movie starring Humphrey Bogart) famously wrote, "Pulp paper never dreamed of posterity."

[8] As I write (2025), there is no known law of physics that out-and-out forbids time machines. But the *engineering* behind actually making a time machine appears to be, shall we say, *of a completely unknown but almost certainly formidable nature.*

Acknowledgments

I have to start by thanking my long-time editor at Springer, Sam Harrison, for his immediate, strong support for the concept of this book. Sam and I have done five previous books together over the years, and I am proud of them all, but this one, our sixth, is at the top of my list. Springer's editorial policy that allows the use of fiction in support of a technical goal is unusual in the scientific publishing world: Springer deserves great praise for that, and I am pleased to be able to enthusiastically offer it here.

Springer Production Editor Ravi Vengadachalam made the process of transforming a collection of WORD files into a beautiful book run smoothly and efficiently. As all book authors know from experience, that is a talent that is worth a *lot*!

Looking far back into the past, I am in enormous debt to all the science fiction editors who published (some of) my stories: Ben Bova, Stanley Schmidt, Reginald Bretnor, Ellen Datlow, T.E.D. Klein, J. E. Pournelle, and Isaac Asimov. For helpful conversations on the business of writing science fiction, I thank Frank Kendig, Ben Bova, and Dick Teresi.

All of the stories in this book were written decades ago (either in my office at the University of New Hampshire or in some long-forgotten airport lounge), but the non-fictional intros and afterwords were all recently written in my favorite writing spot: an old, beat-up leather chair in a quiet corner of Me & Ollie's Coffee shop on Water Street in Exeter, New Hampshire. I offer unbounded kudos to the great staff at Me & Ollie's who kept my writing muse energized with the best coffee anywhere.

And finally, as I always do with a new book, I thank my loving and supportive wife, Patricia Ann. Pat has listened (for 63 years as I write) to all of my "what-if?" ideas and often offered suggestions that made their way into my tales. She didn't hold back when I went off the deep-end with some idea that she declared just a bit *too* crazy, and that was good—a perceptive critic is what every writer should hope for!

October 2025
Exeter, NH, USA

Paul J. Nahin

Competing Interests The author has no competing interests to declare that are relevant to the content of this manuscript.

Contents

1 **Introduction** 1
"Publish and Perish" 10

2 **Far-Out Math** 19
"Mathematical R&D" 26

3 **Interstellar Invasion** 31
"A Quiet, Rainy Afternoon" 34

4 **Revenge** 41
"The Reunion" 43

5 **Invisibility** 51
"Archival Voice" 53

6 **Digital Pictures** 65
"The Last Word" 67

7 **Time Travel** 71
"Old Friends Across Time" 75
"The Invitation" 82

8	**Future War**	97
	"The Man in the Gray Weapons Suit"	99
9	**Topology and the Fourth Dimension**	109
	"Twisters"	113
10	**Alternate Universes**	123
	"The Infinite Plane"	127
11	**National Security**	139
	"Security Blanket"	140
12	**Computers and Society**	151
	"Qualification Test"	152
13	**Far-Out Physics**	163
	"What *Really* Caused the Energy Crisis"	165
14	**Super-Science**	177
	"The Language Clarifier"	179
15	**Artificial Intelligence**	187
	"The Box"	188
16	**Religion**	195
	"A Father's Gift"	197
	"Some Things Just *Have* to Be Done By Hand"	206

1

Introduction

Imagine that you are a visitor to a college campus and are enjoying lunch in the faculty club dining room. By chance you are seated near a table at which two professors, one of mathematics and the other a mathematical physicist, are bringing each other up-to-date on their respective research efforts. You don't mean to listen-in, but they are so excited about their work that you can't help but overhear:

> The equations are just beautiful, Frank, I couldn't believe it! I showed them to my graduate students, and they were blown-away, too.
>
> I know what you mean, Sam. When my original computer plots came out of the printer, they were so unexpected I was absolutely convinced they simply couldn't be right. But later work verified them. The analytical physics is unbelievably pretty!

Sam and Frank euphorically continue along these same expressions of rhapsody all through lunch, and you can't help but smile and (in just your imagination, of course, because you are polite) roll your eyes. "Well, this certainly proves it," you think, "mathematical people *are* a strange lot!"

Your bemused reaction, in fact, is not unusual. After all, most people find that *their* happiest thoughts about mathematics are mostly limited to when the monthly ordeal of balancing the family checkbook results in an actual balance. And when it doesn't balance, most people just say the hell with it and (if the discrepancy isn't too big) simply accept the bank's value and move on.

For a mathematician or a mathematical physicist, however, even a one-cent difference will invariably result in spending maybe *hours* in tracking-down

© The Author(s), under exclusive license to Springer Nature Switzerland AG 2026

P. J. Nahin, *The Art of the "What-If?"*, https://doi.org/10.1007/978-3-032-15618-1_1

where the numbers went wrong, with the eventual discovery of the error resulting in a victorious arm-pump and a shout of "Aha, gotcha!" (behavior almost as ridiculous as when golfers dressed in designer slacks do the same thing after sinking a "brutally difficult" one-foot putt).

Yes, that *is* odd. Who would argue otherwise?

But here's the *really* odd thing—while most people do find that sort of obsession a tad strange, they also are vastly impressed by such absolute dedication to the pursuit of absolute perfection.

Writers of fiction have made good use of this common identification of skill at mathematics with the achievement of superiority (in some sense). What that sense is depends on what the writer is attempting to do in his story, but often it is simply to establish a character as having great intelligence. A prime example of that is when Arthur Conan Doyle (1859–1930) has Sherlock Holmes describe the master criminal, Professor James Moriarty (said by Holmes to be "the Napoleon of Crime" in "The Final Problem"), to Doctor Watson. Watson learns that Moriarty was the author, at age 21, of "a treatise upon the "binomial theorem" which won him the mathematical chair at one of our smaller English universities." And in *The Valley of Fear* we learn, along with Watson, that Moriarty was "the celebrated author of *The Dynamics of an Asteroid*, a book which ascends to such rarefied heights of pure mathematics that it is said that there was no man in the scientific press capable of criticizing it."[1]

After such a promising start, the underlying evil of Moriarty's nature made itself known, and he was forced to resign his chair and move to London where, sitting like a spider at the center of its web, he became the central, monstrous force powering London's criminal world. Holmes confesses to Watson that Moriarty is his intellectual equal and we believe that, without hesitation, because Moriarty's intellectual brilliance has been solidly established via mathematics.

Even in writing as different from a Sherlock Holmes story as you can imagine, like the Harry Potter children's series by J. K. Rowling, we find Harry, in the very first volume (*The Sorcerer's Stone*), invoking his skill at mathematics to defend his honor. When Harry learns his long-dead parents had been wizards (both were graduates of Hogwarts School of Witchcraft and Wizardry) it is because he is visited by Hagrid (Keeper of the Keys at Hogwarts) at the

[1] What could have been in a work with a title like that? Science fiction writer Isaac Asimov (1920–1992) wrote a clever, math-free, speculative essay (with the same title) on that very question, which you can find in the anthology *Beyond Baker Street*, Bobbs-Merrill 1976. There you will learn how Moriarity's fiendishly twisted mind was swept-up-and-away by a bizarre vision of mathematics, *theoretically*, giving one the power to destroy *entire worlds*.

home of his aunt and Uncle. They had taken him in as a baby after his parents died, but had told him nothing of his unusual heritage. Hagrid is shocked at that lapse and angrily growls "Do you mean to tell me that this boy—this boy!—knows nothin' about—about ANYTHING?" To that Harry begs to differ: "I know *some* things. I can, you know, do math and stuff."

And with that single sentence we are assured that Harry is a bright lad, one who will be able to succeed in the magic classes that await him in his later adventures at Hogwarts.

A somewhat different but related use of mathematics is to create a general atmosphere of mystery. A good example of that occurs in the 1897 novel *The Invisible Man* by H. G. Wells (1866–1946). After a bizarre-looking man, wrapped head-to-foot in bandages and wearing heavy clothing and a hat, arrives in a snowstorm at the Coach and Horses Inn, it doesn't take long for the local citizens to grow intensely curious about him. One of them snoops in the stranger's rooms at the Inn after the stranger has gone to bed and, at one point, he "scrutinizes closely ... a sheet of mathematical computation the stranger had left." Who can read that and not wonder just what the invisible man might have been calculating?

The same use of mathematics appears a century later in Dan Brown's 2003 mystery novel *The Da Vinci Code*, whose protagonist is an art historian. In that work there is much talk of *Fibonacci sequences* (in which every term is the sum of the previous two terms) as part of a secret code, and how the number called *phi* that comes from such sequences is of great interest to art history scholars. You can start a Fibonacci sequence with any two positive integers, with the classic example being 1, 1, 2, 3, 5, 8, 13, 21,.... The amazing thing is that no matter how you start a sequence, the ratio of any term to the previous term approaches the *same* value. That is, if x_n is the nth term, then $\lim_{n\to\infty}\frac{x_n}{x_{n-1}}=1.618\ldots$, the irrational number phi that is termed the "golden ratio" and is much admired by art historians. (See the following box for a little more on this.) This is all Brown says about phi, but it is enough for his purposes.

The mere mention of mathematics invariably stirs the imaginations of people, even those whose interest in mathematics as a subject is pretty minimal. Why?

There is probably more than one answer to that question, but what underlies all of them is the unique status of mathematics. In mathematics, once a result is established, *that's it*. Nothing, *absolutely nothing*, will result in having to go back and revise things. You can't say that about engineering, physics, or chemistry. In those fields of study, there are, *daily*, discoveries of how to do or explain things "better" than before. Take physics, for example. Aristotle's view of falling bodies had to be corrected by Galileo, whose work in turn was extended by Newton who was in turn both corrected and extended by Einstein. And it's anybody's guess what will come after Einstein, but the sure bet is that *something will.*

In electrical engineering, to take my own field for another example, the changes have come even faster: in a period of less than 75 years the relay gave way to the vacuum tube which gave way to the transistor which gave way to the integrated circuit which will give way to …? In mathematics, however, what we call the Pythagorean theorem is today just as it was almost 4000 years ago. 1 + 1 = 2 will *always* be 1 + 1 = 2, and 2 will *always* be greater than 1 and smaller than $\sqrt{5}$. These simple but profound observations were clearly stated by Isaac Asimov in the "Introduction" he wrote for the 2nd edition of Carl Boyer's masterpiece *A History of Mathematics* (Wiley 1991): "Now we can see what makes mathematics unique. Only in mathematics is there no significant correction—only extension …Euclid was incomplete, and his work has been extended enormously, but it has not had to be corrected. His theorems are, every one of them, valid to this day." As Edna St. Vincent Millay wrote more than a century ago, in what may be her most famous poem, "Euclid alone has looked on Beauty bare."

Where the "Golden Ratio" in *The Da Vinci Code* Comes From

This is an example of one of the rare times I will do some real math in this book, and that's because all we'll need is high school algebra. A sequence of non-negative integers x_n is a *Fibonacci sequence* if $x_n = x_{n-1} + x_{n-2}$, $n \geq 3$. With the selection of starting values for x_1 and x_2, it is obvious that subsequent terms become large very quickly. Let's suppose they grow *exponentially* (this is an *assumption*, which means I don't have to provide a justification but, rather, we'll just see where it takes us). That is, suppose as n gets "large" that we can write $x_n = Ce^{an}$, where C and a are positive constants. Thus, for large n we can write $Ce^{an} = Ce^{a(n-1)} + Ce^{a(n-2)}$ and, with the obvious cancellations, we have $1 = e^{-a} + e^{-2a}$ or $e^{2a} - e^a - 1 = 0$. This is a quadratic in e^a because writing $y = e^a$ we have $y^2 - y - 1 = 0$ and $a = \ln(y)$. Solving the quadratic, we immediately get $y = \frac{1 \pm \sqrt{5}}{2}$; we reject the negative sign since it is only the *positive* sign that gives a positive value for a. In particular, $a = \ln\left\{\frac{1+\sqrt{5}}{2}\right\}$. Now, calculating the ratio of adjacent terms, we have

$$\lim_{n\to\infty}\frac{x_n}{x_{n-1}} = \frac{Ce^{an}}{Ce^{a(n-1)}} = e^a = e^{\ln\left\{\frac{1+\sqrt{5}}{2}\right\}} = \frac{1+\sqrt{5}}{2} = 1.618\ldots,$$ the Golden Ratio

in Dan Brown's novel. Note that this result depends *in no way* on the starting values for the sequence. The Golden Ratio is interesting to art scholars because it is often claimed that rectangles with side lengths having that ratio are aesthetically the "most pleasing."

For writers of fiction who have a technical bent, the power of mathematics to seize the imaginations of readers is irresistible. The result is often a story of either science fiction or of fantasy,[2] and the distinction between the two possibilities is important to understand. The two great pioneering exponents of these two types of speculative fiction are the Frenchman Jules Verne (1828–1905) and the Englishman H. G. Wells. The two are often lumped together as "writers of science fiction," but in fact their tales are vastly different in nature.

Verne's great works (*Around the World in 80 Days*, *20,000 Leagues Under the Sea*, *From the Earth to the Moon*) are works—with one glaring exception—involving the extension of already-known *engineering* into the near future. Beginning with his 1863 *Five Weeks in a Balloon*, Verne wrote novels that are generally lumped together as his tales of *extraordinary voyages*, and it was with his very next effort in 1864, *Journey to the Center of the Earth*, where he deviated from known science. By 1800, it had been firmly established that the old "hollow Earth" idea was wrong. *Journey* is a great story, yes, but it is fantasy, not science fiction. So, perhaps it is better to say Verne *usually* wrote science fiction.

Wells' great works (*The Time Machine*, *War of the Worlds*, *The Invisible Man*) are, in contrast, works of pure fantasy (or, as they were originally labeled, *scientific romances*). Both Verne and Wells fully understood the distinction between science fiction and fantasy, and each man strongly argued that they were not mere cross-English Channel versions of each other. Their fiction was, instead, as different as day is from night—but one common feature is that both Verne and Wells routinely mentioned mathematics in their stories.

Mathematical fiction does offer a challenge to authors, in that it is potentially open to the criticism of being unemotional, as being writing that offers only relentless logic with no room for human empathy. That view was

[2] An example of this is the hilarious math fantasy "The Devil and Simon Flagg" by Arthur Porges (1915–2006). Published in the August 1954 issue of *The Magazine of Fantasy and Science Fiction*, the story has a math professor engaged in a "deal with the Devil," with the challenge to the Devil being to resolve Fermat's Last Theorem. By the end of the tale, you feel sorry for the Devil! Porges had a Master's degree in mathematics, and the story showcases his technical expertise.

dramatically challenged with the publication of "The Cold Equations" by Tom Godwin (1915–1980). This story, one of the most famous in science fiction, appeared in the August 1954 issue of *Astounding Science Fiction Magazine* (later renamed *Analog Science Fiction/Science Fact*), whose editor John W. Campbell (1910–1971) was instrumental in bringing science fiction out of the "super science, space-opera, ray-guns and pirates-in-spaceships" era of the early pulp magazines. "The Cold Equations" starts with a simple premise that appears to be following the common science fiction formula of "present an apparently unsolvable problem and then, suddenly, have the hero come-up with a fantastically clever solution."

Godwin opens his tale by quickly setting the stage. A small rocket (an Emergency Dispatch Ship, or EDS) is transporting urgently needed medical supplies to an exploration team on a remote planet, lightyears from Earth. The EDS is powered by ordinary rocket fuel, of which EDS carries only a very carefully calculated supply, enough that is *just* sufficient for it to complete a relatively short trip. The major part of a medical emergency mission is performed by the EDS being transported to near the destination planet by a huge interstellar cruiser that can travel vast distances via hyperspace.[3] Once launched from the cruiser, the cruiser returns to hyperspace and the EDS is, with its lone pilot, on its own with its limited fuel supply.

After launch, the story's first sentence is a chilling realization by the EDS pilot:

> He was not alone.

The EDS has a stowaway on-board, a young girl who snuck aboard simply because she wants to visit her brother who is a member of the exploration team. Her presence, however, means the EDS has insufficient fuel to complete its journey, as her unanticipated, extra mass has thrown the fuel calculations out the window. The only solution, dictated by the cold, emotionless equations of mathematical physics, requires that the girl be jettisoned. If she stays, then the EDS can't deliver the urgent medical supplies and the exploration team (and the pilot and the girl, too) will die. If she is jettisoned, then only the girl dies. And so out she goes, to the stunned surprise of *Astounding*'s

[3] Hyperspace is a science fiction concept that allows faster-than-light travel (a necessary requirement for crossing the vast distances of interstellar space in reasonable time), and the idea can be traced back to H. G. Wells' 1895 short story "The Remarkable Case of Davidson's Eyes." Given a space of dimension d, hyperspace (for that given space) has dimension of at least $d + 1$. So, for example, the ordinary three-space we live in is hyperspace for a two-dimensional world (like Edwin Abbott's famous 1884 *Flatland*). In "The Cold Equations," hyperspace is at least four-dimensional. We'll encounter hyperspace again, later in this book.

readers who were expecting a Hollywood-movie resolution equivalent to "end-the-movie-anyway-you want-except-you-can't-kill-the-dog."

The story ends with a line that is even more shocking than is the opening. The once-again lone pilot knows he will never forget the last words of the girl, just before she is ejected: "I didn't do anything to die for … I didn't do anything." Yes, the girl's death seems so unfair, but t*hat*'s the point of the entire story—the universe simply *doesn't "care"* if you are a good person or a bad person. If the math says you have to die, then you must die.

That ending started a crusade by many readers, to show that there is a solution that allows the girl to live. Why not, wondered some readers, jettison instead the pilot's chair, or the cabinets in which the medical supplies are stored? Well, the answer to that and to any other proposed solution is simple: *the result wouldn't have been a memorable story*! Readers who grumble about Godwin's ending apparently don't know that in his original submittal to *Astounding* the girl *is* saved. It was editor Campbell who insisted that the girl had to die, and he rejected revision after revision (in each of which Godwin kept coming-up with ingenious ways to save her) until, finally, Godwin got the message. The result was one of the greatest short stories in science fiction, one that makes readers *think*.

I have, so far, emphasized the linkage between mathematics and science fiction/fantasy and so, for balance, let me switch to a little-known similar linkage between science fiction and physics. In the final days of the Winter of 1891, a curious little note appeared in the letters section of the prestigious British science journal *Nature*. That note's title and opening sentence were, either by intent or by chance, perfect for capturing the attention of all who read them:

> Having had occasion to devise a high-speed whirling machine, I examined the speed at which it might be safe to work, and some of the results surprised me,[4]

Just 4 years later, with the appearance of Wells' *The Time Machine*, that opening sentence could have been used by Wells (when the Time Traveler reveals to dinner guests that he has been experimenting with time travel) if he had simply replaced the words "high-speed whirling" with "time." The author of the note, Oliver Lodge (1851–1940), was a well-known British physicist who held a professorship at University College, Liverpool, and who was a Fellow of the Royal Society. With his note Lodge anticipated, by *almost 80 years*, Larry Niven's classic 1970 novel, *Ringworld*, as well as identifying a

[4] Oliver J. Lodge, "The Flying to Pieces of a Whirling Ring," *Nature*, March 12, 1891, p. 439.

fundamental technical puzzle in the novel's concept. In fact, Lodge had himself been anticipated by a fellow British mathematical physicist, Professor (at Cambridge University) John William Strutt (1842–1919)—better known as Lord Rayleigh—who had considered the same physical situation 15 years earlier, in 1876. Lodge had, in fact, gotten the idea for his whirling machine after a direct communication with Rayleigh.[5]

The subject of Niven's novel, the magnificent Ringworld, is certainly one of the most imaginative and romantic ideas in all of science fiction. Niven asks readers to imagine the following: Take a huge chunk of matter the size (mass and volume) of Jupiter and reshape it into a long, straight strip. If the width of the strip is one million miles from edge-to-edge, and if the strip is 3000 feet thick, then the length of the stripe is 600 million miles. Next, stand the strip upright on one of its longer edges, bend it into a circular shape, and then join the two far ends of the strip together. The result is a two-sided, ribbon-like ring, with a radius of 93 million miles and its inner-surface facing the center of the ring. After doing all that, the next two steps should be (relatively) easy: First, at that center place a star much like Earth's sun, and then second, set the ring to rotating around the star at a speed of 770 miles per second. With that all done, the centrifugal force (think of a merry-go-round) "felt" by beings on the inner surface of the ring is just about equal to the gravity force experienced on the surface of the Earth.

What's the purpose of all this incredible engineering? Well, the artificial habitat called Ringworld has an inner surface area on the ring that is *three million* times the surface area of Earth, and so there is *plenty* of room for everybody! If a country has an assigned area in the shape of a square a million miles on an edge (an area of a *trillion* square miles!) it would be *very* difficult to wage war with other countries, much less to actually invade a neighboring country. Ringworld has room for 600 such countries.

There are numerous technical issues associated with the construction of Ringworld, and Niven discusses many of them in the novel. The most immediate one is that the ring will *disintegrate* (!) long before it can be spun up to 770 miles per second, and this concern was the motivation behind Lodge's note. As he wrote about what in his analyses "surprised" him: "[A] band of

[5] You can find a complete discussion of the underlying mathematical physics of Lodge's whirling machine, including the problem's historical origin with Rayleigh, in my paper "Rayleigh's Rotating Ring," *The Mathematical Intelligencer*, Summer 2017, pp. 72–75. You can read more on just what Lodge was up to with his amazing machine in Bruce Hunt, "Experimenting on the Ether: Oliver J. Lodge and the Great Whirling Machine," *Historical Studies in the Physical and Biological Sciences* 16, no. 1, 1986, pp. 111–134. As Hunt explains, Lodge was attempting to perform a crucial test of a physical theory, and certainly was not thinking of a fictional story (which makes his work all the more interesting, with its implicit hint at *Ringworld* physics).

steel just able to bear a load of 30 tons to the square inch [before breaking] will fly to pieces at a peripheral speed of about 800 feet per second" and so "Hence, a steel girdle around the earth's equator would burst [if it wasn't attached to the Earth]."[6] Lodge was particularly interested in this possibility because his whirling machine used a powerful electric motor to spin massive steel plates at up to 4000 rpm. One of his friends had warned him that if there were any flaws in the plates they could disintegrate and Lodge "would have his head cut off!"

Ringworld's peripheral speed of 770 miles per second (about four million feet per second) *vastly* (obviously) exceeds Lodge's bursting speed of rotating steel or, for that matter, that of *any* known substance! Niven realized that, too, and acknowledges the problem several times in the novel. For example, one character describes the ring as "not gas nor dust, not even asteroidal rock, but a solid band of considerable tensile strength," and then continues with "Consider the tensile strength needed to prevent the structure from disintegrating." One astonishing property of such a ring is that it must be made of "a substance dense enough to stop 40 percent of neutrinos." This is astonishing because neutrinos can easily pass through *lightyears* of lead, while the ring is only 3000 feet thick.

There is nothing known to modern physics that could be a candidate for the ring material, as even the density of nuclear matter (the *ultimate* density stuff, short of the infinity of the singularity that general relativity suggests exists at the center of a black hole!—a mere teaspoon of nuclear matter would weigh a *trillion tons*!) wouldn't do the trick. Such dense matter would, further, have an enormous gravitational field, an effect not mentioned in *Ringworld*, although Niven did use that fact in his earlier short story "Neutron Star" (published in the October 1966 issue of *Worlds of If*). The puzzle of what Ringworld is made of, alone, explains why *Ringworld* is an "imaginative" read.

There are other physical concerns in *Ringworld* besides the question of what the ring is made of, and here are two more. First, the energy required to spin-up a ring with the mass of Jupiter to 770 miles per second is absolutely enormous, equal to the *total yearly* energy output of *thousands* of stars the size of the sun. Second, assuming the ring *can* (somehow) be made and spun-up, there is the issue of *stability*. That is, after the ring has been centered on its star, *any* deviation of the centering of the ring, no matter how slight, tends to *increase* that perturbation and so, eventually, without intervention, the star

[6] See note 4 again for the math/physics behind Lodge's value of 800 feet/s as the speed at which a rotating steel band will disintegrate. At the equator, such a band would be rotating at 25,000 miles (the distance around the Earth) per 24 h. That is, at more than 1000 miles/h which is 1500 feet/s, almost twice Lodge's bursting speed.

and the ring will collide. Niven addressed both of these concerns in the novel's 1980 sequel *The Ringworld Engineers*. It is probably obvious by now that the engineering of Ringworld demands capabilities *far* beyond anything presently known (or even imagined, outside of fiction). That's why *Ringworld* is *science fiction* for many, but is *fantasy* for others.

In this book, then, both science fiction and fantasy will be of interest to us and, to illustrate the difference, what follows is a *science fiction* tale to contrast with the fantasy story that ended the Prelude. It originally appeared in the April 1978 issue of *Analog Science Fiction/Science Fact Magazine* and was inspired by the once fierce debate in the physics world on the possibility of cold fusion. That is, the possibility of achieving nuclear *fusion* (it is fusion that powers the stars, not the chain reactions of nuclear *fission* that are behind "ordinary" atomic bombs and nuclear power reactors) in water at room temperature. I think cold fusion vastly unlikely, because of the energies involved in a fusion reaction, but still—*what if*? *Cold fusion would change the world*!

The result of that wondering was "Publish and Perish." Its format is what is called *epistolary*, a story told as a series of letters. The story also treats, as a secondary theme, the matter of security clearances. Before I became an academic, I worked for some years in the Southern California aerospace business, and then later at two military think tanks in Washington, DC, in positions that required security clearances (first Secret and then Top Secret). My experiences with what the American government considered "secret" greatly influenced the tone of "Publish and Perish."

"Publish and Perish"

March 17, 1977

Mr. Thomas W. Starr
3613 Laguna Avenue
La Mesa, California 92041

Dear Mr. Starr,

It is with pleasure that I welcome you to the Faculty of the California Technological Institute. Dean Johnson has informed me that you have accepted the appointment of Instructor in Physics. Your title will change to Assistant Professor upon completion of your doctoral studies.

We look forward to seeing you in September.

Sincerely,
Dr. W. Allen Smith
Office of the President

May 3, 1977

Mr. Thomas W. Starr
3613 Laguna Avenue
La Mesa, California 92041
Dear Tom,

President Smith has asked me to write to you about an issue we wish to have clear, before you arrive on campus in September. Your appointment is for two years (renewable), but you must receive your PhD by December 1, on your second academic year with us. Otherwise, because of the Institute's bylaws, we would be unable to recommend your reappointment to the Trustees.

Sincerely
Peter V. Johnson
Office of the Dean

May 8, 1977

Dr. Peter V. Johnson
Dean of Engineering and Science
California Technological Institute
Claremont, California 91711
Dear Dean Johnson,

The completion of my doctorate by December 1, 1978 will be no problem. I have talked this matter over with my dissertation advisor, Professor B. B. Abernathy at San Diego Tech. He assures me that there will be no difficulty. He is even joking about there being a Nobel Prize in it for the two of us!

Cordially,
Thomas W. Starr

September 9, 1977

Mr. Thomas W. Starr
1713 12th Street
Claremont, California 91711
Dear Tom,

Sorry we missed our regular meeting last week, but I couldn't skip the review briefing on my grant at the Pentagon (but the thought of the agony of the red-eye flight to DC *almost* made it worthwhile to cancel out). While losing my way wandering around the Puzzle Palace, I happened to mention some of your most recent results toward inducing nuclear fusion in water, and it caused quite a stir. If you get tired of teaching in Claremont,[7] give the Civil

[7] There is no California Technological Institute in Claremont, but there is Harvey Mudd College (where I once taught), and there is the California Institute of Technology in Pasadena (where I got my Master's

Service a thought — you can't believe some of the nitwits with GS-12 and 13 ratings they've got back here. A good man like you would get snapped right up, and it certainly beats what they pay young, new college teachers.

See you next week, and we will discuss that first draft of your thesis.

Regards,
Bertram B. Abernathy, PhD
Professor of Physics
San Diego Technical University

---

14 October 1977

Mr, Thomas W. Starr
Physics Department
California Technological Institute
Claremont, California 91711

Dear Mr. Starr,

It has recently come to our attention that you have been pursuing innovative concepts on the possibility of inducing nuclear fusion in water. Your doctoral research support is funded through an Army Office of Scientific Research grant to Professor B. B. Abernathy and, as you know, we retain the right to request periodic reviews of research supported by us.

Professor Abernathy has informed us that you are now writing a thesis for open publication, based on your work. Please send three (3) copies of your draft to:

Colonel Andrew Bobble
Chief, Nuclear Security Review Office (Army)
The Pentagon
Washington, DC 20310

Sincerely,
Patricia Adams
Administrative Assistant
Nuclear Security Review Office (Army)

---

7 January 1978

Mr, Thomas W. Starr
Physics Department
California Technological Institute
Claremont, California 91711

---

degree). CTI is a composite of the two schools. San Diego Technical University exists only in my imagination.

Dear Mr. Starr,

After a careful review of the material you recently sent to us for review, we have classified it. Please forward all additional copies of your thesis drafts, plus any other related documents, within ten days, by registered mail, in a sealed envelope within a sealed envelope.

Sincerely,
Colonel Andrew Bobble
Chief. Nuclear Security Review Office (Army)
The Pentagon

January 11, 1978

Colonel Andrew Bobble
Chief, Nuclear Security Review Office (Army)
The Pentagon
Washington, DC 20310

Dear Colonel Bobble,

I have read your letter to me of January 7, and I am at a loss to understand what you mean by 'classifying' my PhD dissertation. I have no security clearance, and at no time have I had access to classified information.

I am sure that any little details in my writing that might cause some concern by your office will be easy for me to work around. If you will send me a list of the particular issues in question, I will be glad to take them into consideration as I finish up my writing.

Sincerely,
Thomas W. Starr

---

20 January 1978

Mr, Thomas W. Starr
Physics Department
California Technological Institute
Claremont, California 91711

Dear Mr. Starr,

This letter is to inform you that there is no appeal from our decision to classify the draft material you recently sent us. It is our final conclusion that there is no possible way to rewrite this material to eliminate the possibility of disclosing information vital to the national security of the United States. We can not transmit the list requested in your letter because such a list would be classified and you have no clearance.

To discuss this matter further, it will be necessary for you to obtain a clearance from the Defense Industrial Security Corporation (at your personal expense), and to travel to Washington to meet with our staff. Even if you decide to do this, we must receive all information still in your possession.

Sincerely,
Colonel Andrew Bobble
Chief, Nuclear Security Review Office (Army)
The Pentagon

January 25, 1978

Colonel Andrew Bobble
Chief, Nuclear Security Review Office (Army)
The Pentagon
Washington, DC 20310
Dear Colonel Bobble

I can't believe this! You are destroying my career with all this Catch-22 crap about national security leaks that you can't tell me about because I don't have a clearance. I have looked into getting a clearance, too. The DISCO investigation fee is $4000. I haven't got forty bucks.

How did you get to be a Colonel? Thinking up stupid things like this? Well, you can go to hell! What I think up on my time, with my brain, is none of the Pentagon's damn business. I will finish my writing, publish it, and the Army can go screw itself.

Sincerely,
Thomas W. Starr

February 1, 1978

Mr, Thomas W. Starr
Physics Department
California Technological Institute
Claremont, California 91711
Dear Mr. Starr,

In response to your letter of 25 January, enclosed is a copy of Title 18 of the United States Espionage Act. Release of classified information is a felony offense, punishable by up to ten years in prison, or up to a $10,000 fine, or both.

We are instructing University Microfilms not to produce microfilm xerographic copies of any unauthorized thesis you attempt to submit to them. In addition, we have notified all domestic and international journals of theoretical and applied physics and/or chemistry that publication of papers by you, without prior release from us, may constitute a security violation.

Please submit all documents on your research to us, as requested earlier, postmarked no later than 15 February 1978.

Sincerely,
Colonel Andrew Bobble
Chief, Nuclear Security Review Office (Army)
The Pentagon

February 10, 1978

Colonel Andrew Bobble
Chief, Nuclear Security Review Office (Army)
The Pentagon
Washington, DC 20310

Dear Andy,

Enclosed are all of the documents you requested from Tom Starr. I am taking care of all these details of transferring the water fusion work to secure, classified areas as Tom is in no shape, emotionally, to do it himself.

I know you understand the reason for his recent intemperate letter to you. He has even cut off his contacts with me, but I think he will come around in time. I am confident of his ultimate discretion and loyalty.

I am pretty sure I can handle the new classified work on water fusion for your office, but we should discuss contract funding levels on my trip to Washington next month. Take me to lunch at the Sans Souci[8] and tell me how much you can give me!

Regards,
Bert Abernathy

December 2, 1978

Mr. Thomas W. Starr
1713 12th Street
Claremont, California 917111

Dear Tom,

I write this letter with regret. You are a talented teacher, and I believe that with time you will become an outstanding member of the academic community. Still, though the circumstances of your doctoral dissertation difficulties were beyond your control, I can not recommend the continuation of your contract beyond June 30, 1979. The Institute bylaws are

[8] Sans Souci was, in the 1970s, *the* trendy eatery for publicity-conscious VIPs in Washington, DC. It went out of business in 1983 and was replaced by a near-by McDonalds.

most specific, and the failure to obtain your PhD by December 1 leaves no room for exception or waiver.

I will, if you wish, do what I can to aid you in seeking a new position for next year.

Sincerely
Peter V. Johnson, PhD
Office of the Dean

July 15, 1979

Dr. Peter V. Johnson
Dean of Engineering and Science
California Technological Institute
Claremont, California 91711

Dear Dean Johnson,

I am writing to thank you for your help in getting me a teaching job for the coming semester. Teaching freshman physics and chemistry at Contra Costa Junior College is going to be a change for me, but without a doctorate I guess I am lucky to have that—at least I won't be pumping gas or hacking a cab. I hope I can find something for the second semester.

The nature of my work in water fusion is such that the lab facilities I saw on my interview at Contra Costa will let me continue. What the Army doesn't know won't hurt me!

Thanks, again, for your help.
Regards,
Thomas W. Starr

December 13, 1979

Mr. Thomas W. Starr
Science Department
Contra Costa Junior College
Walnut Creek, California 94596

Dear Mr. Starr,

I am pleased to inform you of the acceptance of your paper "An interesting classroom demonstration of power from water." It will appear in our issue of February 24, 1980.

Quite frankly, we were astounded when we duplicated the techniques described your paper. We would be interested in seeing a second paper which elaborates, mathematically, on the specific chemical and physical processes of your demonstration, as we believe the present paper will attract considerable attention.

We wish you well on your new post at the South Australian Boy's Military Prep School, and the galley proofs of the paper will be sent to you there.

Cordially,
Peterson S. Day
Editor, *Review of High School Experimental Science*

****

You might now be wondering about what happened next to Tom Starr? Would US agents track him down at some remote Australian high school and arrest him? Would they do that if Tom received the Nobel prize in physics (a sure thing, I think, for the discoverer of how to achieve nuclear fusion in room temperature water!)? Would the Army, instead, try to spin the narrative to how it was *Army* money that had unlocked cheap, unlimited energy for the world? *Analog*'s editor convinced me that to try to properly treat such questions would turn the story from a *short* one into something that would have to be much longer, a task which I had no desire to pursue. So—this book is all about *imaginative* fiction and I therefore leave the answers to those questions up to *you*.

Finally, to end this introductory chapter, let me point out that science fiction (with its pulps) was *not* the only genre with a literature exploring "what-if?" futures. For example, I mentioned back in note 3 of the Prelude the "magical" nature of the development of broadcast radio in the early 1920s. Those early days of radio generated the same excitement that the early days of the internet did decades later, and a large number of young people were attracted by the seemingly endless possibilities that electronics appeared to offer (many of those youngsters were also devoted readers of the science fiction pulps). The appearance of radio prompted the publication of magazines for *its* legions of enthusiasts, one of which was the *Citizens Radio Call Book and Scientific Digest*, which carried news of the latest developments in radio. In Fig. 1.1 you see the cover art for the September 1928 issue, a painting which implicitly asks a "what-if?" question: "radio transmits sound, but what if it could also transmit pictures—in color?"

Fig. 1.1 In 1928, color television was but a scientific dream, one not *quite* as outrageous as asking for a time machine—but pretty close!

2

Far-Out Math

In this chapter, we will begin with a short story that was written for one reason only: to tell a somewhat sophomoric (in my opinion) sexual "joke." Despite that, what makes this story significant for us is that the "joke" is powered "behind the scenes" by a very interesting *math* problem, a problem that I think the author actually failed to appreciate. So, in that sense, Fredric Brown's "Expedition" (published in the February 1957 issue of *The Magazine of Fantasy and Science Fiction*) is an *accidental success*. Well, accident or not, I think you will find it well worth the time for a close look.

Fredric Brown (1906–1972) was an ingenious writer, one well-known in the story-telling world as a master at the short-short story (less than, say, 1000 words) with a sudden, surprise ending. "Expedition" is a good example of his short-short work and, while I think the "joke" a bit lame, you still simply have to admire the sheer cleverness of the story. Brown's personal life shows no training or interest in mathematics (his "day job" was as a proof-reader for a newspaper) and that is why I think he didn't appreciate the serious mathematical nature of his tale. For him, the "joke" was the entire point.

The premise of the story is straightforward. The tale is told through a narrator (a history professor) who is lecturing a class of students on the details of the first colonization mission to Mars, a mission that happened some years earlier. The mission was designed to have a crew of 30, with the central administrative question being the determination of how many of the 30 should be men (and so, of course, how many would be women). The men and women were to be selected from the two Space Cadet Schools (500 men at the all-male school, and 100 women at the all-female school). There were three proposals on how to it, with each having its supporters.

© The Author(s), under exclusive license to Springer Nature Switzerland AG 2026

P. J. Nahin, *The Art of the "What-If?"*, https://doi.org/10.1007/978-3-032-15618-1_2

1. An equal number of each gender.
2. 25 men and 5 women
3. All men, no women.

Proposal (1) seems to be the one that first comes to mind for many people, while with proposal (2) we see the opening gambit of the "joke."

The decision of which proposal to go with proved to be a political minefield and so, the Director of Space Travel being a political animal, he decided to *not* make the choice but rather to let the workings of probability select the crew. That is, the lucky 30 were chosen from the 600 men and women at the two Space Cadet Schools by random lot. The result shocked everybody: 29 women and just *one* man! (With that, you can almost hear the rest of the "joke" clicking into place.) As Brown states in the story,

> By the law of averages the ratio of winners should have been five men to one woman. However, the law of averages[1] does not always work out [for any particular case].

To appease male egos, the lone man (named Moxon) is named Captain of the crew.

With the crew selected, the rocket blasts off for Mars and arrives okay. And then, at last, the final bit of the "joke" is revealed: exactly 9 months and 2 days after the rocket left Earth, a second rocket arrives to find that every woman in

[1] As the great Princeton University probability expert William Feller wrote in his classic text *An Introduction to Probability Theory and Its Applications* (3rd edition), Wiley 1968, p. 88, concerning this "law," he attributed it "to our faulty intuition and to our being exposed to too many vague references to a mysterious 'law of averages.' There is, in fact, no such law." For the sake of the story, however, we will go along with Brown on this point and ignore Professor Feller.

the first crew has given birth to a baby (one had twins!) and so the lone man, the Captain, proudly earns the sobriquet of "Mighty Moxon."

Uh-huh. Well, that *was* the *1950s*.

As Brown no doubt hoped, the revelation of mass pregnancy earned appreciative chuckles from many of his mostly young, male readers.[2] As you read that sentence, don't think I am unfairly targeting men, because in the 1930s, 1940s, and 1950s, science fiction was, in fact, not the only fictional genre published for a mass audience on a regular basis. Romance, mystery, horror, crime, and Western stories, too, were pumped-out like widgets on a production-line, and lots of it was pretty awful. As one observer of pulp romance fiction, for example, bluntly put it, the heroes and heroines in those tales often displayed the "mental equipment of a banana split." The strong implication was that the same description applied to the female readers, as well.[3]

But that description would, I believe, generally *not* apply to the young male audience of science fiction. That says nothing, however, about that audience's emotional maturity. In the pulp science fiction of the 1920s through the 1950s, a common theme was that of grotesque, salivating, bug-eyed monsters clutching screaming, half-naked Earth women in claws, tentacles, or some other disgusting appendage. The lurid, brightly colored covers of the pulp science fiction magazines, illustrating such quasi-erotic goings-on, were often simply cheap substitutes for *Playboy*-like centerfolds. How could immature teenage boys (even those who were smart enough to be good at math and physics!) possibly resist such stuff?[4]

Well, lots of them couldn't but, fortunately (for both readers and science fiction, itself), after the 1950s times changed for the better!

Now, at last, we come to the math problem inherent in "Expedition." We are told in the story that the "law of averages" has failed—but not a word is given to describe by how much the crew lottery has deviated from the "expected." That is, the question a mathematician or physicist or engineer would almost reflexively ask is: what is the probability that a lottery involving

---

[2] As one critic put it, science fiction "is a genre noted for stereotyping men as Messianic heroes." See Beverly Friend, "Virgin Territory: the bonds and boundaries of women in science fiction," in *Many Futures, Many Worlds: theme and form in science fiction*, Kent State University Press 1977, pp. 140–163.

[3] Margaret MacMullen, "Pulps and Confessions," *Harper's Monthly Magazine*, June 1937, pp. 94–102. By the 1930s, all the pulp magazines were lumped together and routinely dismissed (by readers of "high class, slick paper journals") as "magazines for morons."

[4] To be fair, many of the cover illustrations of the science fiction pulp magazines were artworks of great talent. See the encyclopedic (nearly 800 oversized, glossy pages!) *Art of Imagination: 20th Century Visions of Science Fiction, Horror, and Fantasy*, Collectors Press 2002. Also excellent in its presentation is Robert Lesser's *Pulp Art: Original Cover Paintings for the Great American Pulp Magazines*, Gramercy Books 1997 (for science fiction art, in particular, see pp. 26–47).

500 men and 100 women will end with 29 women and 1 man for the selected crew of 30? Not to give too much away—it is really, *really* small! *So* small, in fact, as to be far and away beyond being simply unlikely. To compute that probability is actually "just" an exercise in counting (no advanced math is required) with the result[5] being that the probability is 1.55×10^{-23}. That's 0.000…0155, where there are 22 zeros to the right of the decimal point before we get to the 155.

The lottery of "Expedition" is a great math problem to show students of probability (the story outcome is *extremely* unlikely, but it isn't impossible), and I did that myself a few times. The idea of embedding a math puzzle in a fictional tale is not a bad idea and, indeed, such tales appeared in fiction long before Brown's story.

One such story was written by Edward Page Mitchell (1852–1927), an American author of science fiction long before the appearance of the pulp magazines; his stories were often published in newspapers. His "The Tachypomp" was published not in a newspaper but in the March 1874 issue of *Scribner's Monthly*, in which the hero is challenged with the task of devising a physical way to obtain *infinite* speed. The clever resolution of this challenge was based on having a series of piled-upon-each-other-trains, with each train riding on rails laid on the train immediately below it. In this way, the individual train speeds were imagined to add and so, as the train pile increased so did the speed of the upper-most train. This may have made some sort of sense in 1874, but, of course, Einstein's special theory of relativity (1905) did it in. Even in 1874, the idea seemed a bit much even to the author and Mitchell has the story end as having been just a dream.[6]

Another "math-in-a-story" appeared in the April 1927 issue of the pulp *Amazing Stories*, written by Harry Stephen Keeler (1890–1967), today a little-known writer of science fiction but one who was very prolific in his time. His "John Jones's Dollar" is essentially an essay, in fictional form, on how interest accumulates in a bank account (opened in 1921 with one dollar) that pays 3% interest, compounded annually. The zinger in the story is that a thousand years later, in the year 2921, the account has the value $\$1.03^{1000}$, a number in excess of six *trillion* dollars, which (the story says) exceeds the value of the

[5] For computational details, see my book *In Praise of Simple Physics*, Princeton 2017, pp. 3–4. To carry the calculations to completion does require the use of a mathematical formula called *Stirling's asymptotic approximation for factorials* (which I simply quote without elaboration in *Simple Physics*), but you can find a derivation and some additional discussion of it (at the level of high school AP-calculus) in my book *The Probability Integral*, Springer 2023, pp. 136–144.

[6] It is interesting to note that Mitchell's story anticipated, by a century, the speculation in theoretical physics of the (so far) hypothetical, faster-than-light particle called the *tachyon*. As explained in the story, the name *Tachypomp* comes from *Tachu* (quickly) and *prepompa* (to send).

entire Solar System! That may well have been thought to be outrageous in 1927 but, of course, today (for better or worse) we find it quite the opposite: as I write this, at the end of 2024, the national debt of America, *alone*, exceeds 36 trillion dollars!

Can we find more effective "outrageous math" stories in the literature? I think so, and I'll start the next part of this chapter by taking you through a better (but still flawed) attempt at "an outrageous math issue in a story." In "The Tenth Question" (*All-Story Weekly*, December 18, 1915) by the American writer George Allen England (1877–1936), we read of a madman who bears a bitter grudge against all physicians. As he explains, it's because "in 1908 … a surgeon did me a tremendous, irreparable wrong, not through malice, but through ignorance, stupidity and mis-judgement … I determined to devote my life to the testing of the intelligence of medical practitioners [and 'removing' those who fail an intelligence test]."

To do that, the madman lures doctors to his home by claiming he is ill, where he then imprisons them in a cage. He will set his prisoner free *only if* he passes the intelligence test. As the madman explains the test to his latest subject, "I have in mind a certain thing … I will give you [the opportunity to ask me] ten questions [one a day, for ten days] to discover [the thing I am thinking of]." If the prisoner fails to successfully determine what the madman is thinking of, the doctor will be killed.

After a detailed presentation of the doctor's growing mental despair at the awful predicament he is in, he is at last down to his last question. The doctor has arrived at the point where he can summarize what he has learned from his first nine questions, as follows: "You are thinking of a useful mathematical abstraction [*that is not a number*]." When the madman says the doctor has it right, the doctor asks his final question:

"Is it—*zero*?"

And *yes*!, *Holy Cow*!, that *is* what the madman is thinking of, and so of course the madman becomes even crazier than he has been, with new, boiling-hot, raw anger at the doctor for having passed the test and …. I'll stop here because the tale now quickly comes to a melodramatic conclusion in which the tables are gratifyingly turned on the madman. What is troubling about this story, for a mathematical person, is that it is of course mathematical nonsense. That's because.

Zero is a number.

What *would* have made sense would have been to have the doctor ask.

"Is it—*infinity*?" because infinity is consistent with *both* requirements (of being a useful mathematical abstraction *and* that of *not* being a number—see the following box for more on that last point). England's imagination failed him at just the point when he needed it most.

Why Infinity Is *NOT* a Number

For mathematicians, infinity is a *concept*, not a number. To see that infinity does not behave like a number, consider this little story, usually attributed to the German mathematician David Hilbert (1862–1943) who was a great admirer of the Russian-born German mathematician Georg Cantor (1845–1918), who pioneered the mathematics of infinite sets (see my book *The Logician and the Engineer*, Princeton 2017). Imagine a hotel with a finite number of rooms (say, 1000), with every one of them occupied. If someone new then arrives at this hotel, he will obviously be turned away as there are simply no rooms left. Now suppose the hotel has an infinite number of rooms, with every one of them occupied. But now if someone new arrives and asks for a room, they *can* be accommodated. Here's how. Just move the person in room n to room $n + 1$. This results in everyone already in the hotel having a room, *and* room 1 now being empty. *That's* the room for the new person. The hotel with an infinity of rooms has not behaved as did the hotel with a finite number of rooms. To really drive this astounding result home, suppose that instead of *one* new person showing up, an *infinite* number of new people arrive. Amazingly, they can all be accommodated, too! Here's how. Just move the person in room n to room $2n$. This results in everyone already in the hotel now being in one of the infinity of *even* numbered rooms, *and* with *all* the *odd*-numbered rooms (also an infinity) now empty. Those are the rooms for all the new people.

Now, before you can ask why I have taken you through England's logically failed story, here's the answer. Or, at least, here is what I admit is a bit of speculation. England's story was read by a teenager, Stanley Weinbaum (1902–1935), who recognized the math flaw in it. Weinbaum, who before his early death went on to write some greatly admired science fiction,[7] decided at some point that while England had a good *idea*, he completely dropped the ball with his ending. So, he decided to rewrite the tale with the title "Real and Imaginary" (which appeared under a different title—more on that soon) in the December 1936 issue of *Thrilling Wonder Stories* (the recently renamed *Wonder Stories*). Weinbaum shifted the story from a madman with a grudge against doctors to a madman with a grudge against mathematicians, but left the gist of the story as in England's.

Weinbaum's version opens with the madman as a chemist[8] who was horribly mangled when an experiment went wrong. He blames the mathematician

[7] For example, "A Martian Odyssey," published in the July 1934 issue of *Wonder Stories*, which was the first science fiction story to treat an alien being as more than a bug-eyed monster.

[8] Weinbaum started his college years not as a math major, but rather in chemical engineering.

who erroneously did the preliminary calculations for the experiment and so, in revenge against all mathematicians, he lures a math professor to his home and challenges him (at gun point!) to figure out what "numerical expression" he (the madman) is thinking of by asking no more than ten questions. Weinbaum's story answers that question in a most imaginative way.

The professor's first question is whether the expression is real or imaginary (recall Weinbaum's original title) and the answer is "either." The professor thinks that means the expression must be zero and is shocked to be told he is wrong. Another question is "What is the expression equal to?" and the madman replies with the even more shocking "anything." Eventually the professor figures-out that the answer is $\infty - \infty$ (infinity minus infinity).

Weinbaum's rewrite isn't without its math stumbles, too. At one point, it is stated the standard math symbol for $\sqrt{-1}$ is e (not so, as the standard symbol for the square root of minus one is i, while e is the standard symbol for the base of the natural logarithms), but this seems, to me, a pretty minor objection. Slightly more worrisome is claiming that $\infty - \infty$ is a *numerical* expression but, even that I think probably gets a pass. In fact, I think having $\infty - \infty$ appear in an early 1930s science fiction tale was pretty impressive, as physicists were already running into just such bizarre results in their studies of quantum theory. That is, when trying to theoretically calculate the value of some physical quantity, one with an experimentally known finite value, the equations often reduced to the difference between two expressions, each of which individually was infinite. What could that mean? and it is still a question physicists are grappling with today.

Weinbaum's story appeared posthumously, and the editorial introduction thanked Weinbaum's widow for sending it to *Thrilling Wonder Stories*. The published title was "The Brink of Infinity," and an illustration for the story showed the math professor being pushed off a cliff by the madman into a "sea of grinning mathematical monsters." It isn't clear who changed the title from Weinbaum's original "Real and Imaginary," but whoever did that apparently thought having the word *infinity* in the title was the better fit with the "surprise" solution to the madman's puzzle.

England died the year the story was published, but *Thrilling Wonder Stories* made no mention of "The Tenth Question" from 20 years before, which might seem to be a questionable slight. It may well be that Weinbaum, himself, wrote his version strictly as a private exercise, with no intention to actually seek publication. With the nearly simultaneous deaths of both authors, however, and with Weinbaum's growing visibility in the science fiction world, it was just too tempting, for both the widow and the magazine, to leave "The Brink of Infinity" lying unsold in a desk drawer.

Fig. 2.1 What-If? Questions in "everyday" life. (By permission of www.CartoonStock.com)

To write a "far-out" math story is probably on the "to-do" list of most science fiction writers, and I was no exception. The result was "Mathematical R&D," which appeared in the January 1979 issue of the professional journal *Transactions of the Aerospace and Electronic Systems Society*, published by the American Institute of Electrical and Electronic Engineers (the IEEE). Besides its obviously outrageous math, there is a more serious point to the piece that I will elaborate on at the end of the tale. But first, before we get serious, let's not be (Fig. 2.1).

"Mathematical R&D"

The lecture hall was packed, and the air buzzed with whispered excitement. Wild speculations on what was about to happen flew from mouth to ear. The great, Full Professor of Mathematics, Dr. Oliver Osgood, head researcher on five simultaneous government grants, was to speak that day at the graduate seminar on his newest, most astonishing discovery.

Not since von Neumann proved the Fundamental Theorem of Games, not even since Fermat scribbled his Last Theorem in a margin, nay, not since Archimedes bounded pi, had such a penetrating intellect laid bare the most intimate details of the Queen of the Sciences. Professor Osgood's insight had once again cast bright light into a hitherto dark corner of mathematics. They were all here to learn what new, incredible secret was now in humanity's possession.

The door burst open (one could almost hear trumpets), and the Great Man strode to the podium. Tall, thin, head awash in a mass of white, unruly hair, eyebrows thickly bushy, he placed a single sheet of paper on the wooden surface before him. He stared at it intently. Impressed, the audience exploded into wild applause. He acknowledged the acclaim with a boyish (yet mature) smile, then held his arms high to hush the excited crowd.

"Thank you, thank you. You are generous, most kind, indeed. It gives me enormous pleasure to be able to share with you the latest fruits of my research. The Government, which sponsors my work, has agreed to the seminar being unclassified, and I wish to thank those enlightened people for the wisdom of their decision."

"Now, let me get right to it, with no further delay. I have found a new number!"

The audience stared, dumbfounded, at Professor Osgood. This was just incredible.

"Yes, between two and three, somewhat closer to the two than the three, is a previously unsuspected number. Like a lost penny it has remained hidden through the centuries among the rationals, secreted away beneath the irrationals, but most deeply covered by the transcendentals. Yet I have found it! With the help, of course, of 43 graduate students, and hundreds of hours on the university computer system. I thank them all."

A mere associate professor jumped to his feet. "But Professor, this is absolutely marvelous! What *is* this number? Will you write it on the blackboard for us?" A hundred pencils leaped into the hands of the audience, quivering in excited anticipation of recording the new number.

Professor Osgood shook his head in a charmingly rueful way. "No, I'm sorry, I can't. As you all know, there are two kinds of mathematical proof. The first kind, a constructive proof, would give us the actual value of the number. But my proof is of the second kind, an existence proof. It shows the number exists, but it doesn't tell us *where,* exactly, it is. But we are working on that! The Government has graciously allocated ten million dollars more for my research in the coming fiscal year."

The audience thundered its approval. Those were big time bucks! There would be summer research money for all!

"Well then, Professor, can you tell us more about your proof?"

"Sorry. That's classified DEEP SECRET. Besides myself, only the Secretary of Defense, the Chairman of the Joint Chiefs of Staff, and the President and First Lady have seen it. I assure you the secret is safe with *them*. I *can* tell you it is based on the fact that three is odd and two is even, but beyond that, I'm sure you understand my reluctance to say more."

The mere associate professor nodded his head to show his understanding, and asked, "Tell us, Professor, does that mean this new number has *military* significance?"

Professor Osgood smiled good naturedly and replied. "We're sure of it. The integrity of our NATO forces in Europe, of our nuclear submarine fleet, of our SAC bases, of our ICBMs, all will be immensely improved by knowledge of this number." He leaned forward toward the spellbound audience.

"I will tell you this much. If the new number is squared, then added to itself, and the result multiplied by the cube root of twenty-nine, the theory indicates we *should* have the absolute, total power to …" Before Professor Osgood can finish, a short, fat, heavily sweating man pushed forward from the rear of the room, knocking people aside like bowling pins as he thundered down the aisle. "No, no, Professor, that's classified DEEP SECRET, too! You must say no more!"

Professor Osgood smiled gratefully at the excited man. "Ladies and Gentlemen, my apologies for this lapse. This is Doctor Thigbit Stanley, IDA—Institute for Diabolical Analyses, my contract monitor—and I thank him for setting me straight." The sweaty, almost dripping Doctor Thigbit sat down with a plop and a wet squish, relieved at having once again avoided a major leak.

The Professor continued. "Our work for the coming year has the highest urgency. Naturally, we *must* beat the Russians to this new number. The whole strategic balance of technological parity could be unbalanced if we lose this race. There *must* not be a Number Gap!" He paused for a dramatic moment as his eyes blazed, his brow wrinkled, and his fists clenched. He cut a mighty impressive figure, he sure did. The room burst into light from the flash camera of the local news photographer. What a picture it would make on the front page of the late edition!

"And finally, ladies and gentlemen, knowledge of this new number will give us an additional bargaining chip at the upcoming arms control talks! The very future of East-West détente may hinge on our work over the next twelve months!" Professor Osgood then picked up the single sheet of paper before him (having by now decided to go along with his broker's recommendation to dump soybean futures and to buy Acme Gaskets), and strode from the room.

The lecture hall rocked with the thunderous applause of the audience. They knew the worth of what they had heard. With scientists like Professor Osgood, the Free World would never have to worry.

Because, as everybody knew, somewhere in the Soviet Union [this piece was published more than a decade before the Soviet Union collapsed] the Russians probably had a Professor Osgood, too.

Years after "Mathematical R&D" was published, it became an entry on a website maintained by Professor Alex Kasman, in the mathematics department at the College of Charleston, South Carolina. That site is devoted to the appearances of mathematics in all fictional genres, not just science fiction. Professor Kasman made the comment, when listing the story, that it was making fun of mathematical research, which was not my intention at all. I wrote to Professor Kasman about this, and he agreed to add my response to the website: "This story was a spoof on the Cold War (which in 1979 was roaring hot and heavy). The worry (in the story) about a 'numbers gap' was a take-off on the then much talked-about 'missile gap.'[9] I simply used the unsuspected number between 2 and 3, found via a computer search, as a device that would be [so I thought] immediately understandable by all of any technical persuasion. I wasn't attacking mathematics research, but rather those who would profit from dubious 'military paper-pushing'."

[9] There were many people, some quite senior in the strategic weapons culture, who argued that if America had "just" 5000 nuclear missiles (*each* armed with *multiple*, independently targeted 170-kiloton warheads), while Russia had 6000, America would be in grave danger. That was the "missile gap." The "missile-gap" people argued that if the other side had 6000 missiles, then we should have 7000! You can read about the often-bizarre debates, at the highest levels of American politics, on how 'best' to blow-up the world in Fred Kaplan's *The Bomb: Presidents, Generals, and the Secret History of Nuclear War*, Simon & Schuster 2020. The detonation of just *one* 170-kiloton warhead (at least 9 times more powerful than the Hiroshima/Nagasaki bombs) over a major city would be an event of unimaginable horror.

3

Interstellar Invasion

As mentioned back in note 7 of the Prelude, science fiction has a long history (over a century as I write) of "publication for the masses." Even earlier, however, is a related genre that while "different" is not actually entirely different from science fiction: *horror*. That one word says it all. A horror story has just one goal, and that is to scare the hell out of you. At first that may seem odd, but it appears that human nature is such that most people *enjoy* such reading and, in fact, it seems the more terrifying the tale the better (just think of the immense popularity of the stories by Stephen King).

Horror fiction has been around for a long time. The science fiction pulps are dated from the early twentieth century, but horror and terror "pulps" can be traced back even further, to the pre-Victorian era tales that appeared in the "penny dreadful" and "shilling shocker" chapbooks (small, cheap-paper pamphlets) of the early nineteenth century. A classic, indeed, famous, well-known example of such a work is *Frankenstein* by Mary Shelley (1797–1851), which appeared in 1818. Even earlier was *The Mask* in 1796 by the English novelist Matthew Lewis (1775–1818), a work described by one critic as "A mass of murder, outrage and indecency," in a review *guaranteed* to attract a certain class of readers![1]

Deviating from the traditional theme of murder and bloodshed, a later form that horror took when appearing in early science fiction was as "an

[1] In note 3 of Chap. 2, I wrote that the general view of the pulps, held by 'sophisticated' readers, was that they were "magazines for morons." To be even more direct with this snobby attitude, sophisticates thought of the readers of the *horror* pulps, in particular, as somehow being even worse: *they* were, in addition to being morons, mere peasants who had nevertheless managed in some mysterious way to learn how to read by laboriously sounding-out the words as they slowly dragged a dirty fingernail beneath the sentences.

© The Author(s), under exclusive license to Springer Nature Switzerland AG 2026

P. J. Nahin, *The Art of the "What-If?"*, https://doi.org/10.1007/978-3-032-15618-1_3

invasion of Earth by creatures with bad intentions from outer space." With modern astronomy's understanding of the sheer immensity of the universe, the term "from outer space" is today taken to mean from anywhere in a volume of space so large that it contains 100 billion galaxies, *each* with 100 billion stars. That so-called *interstellar space* provides writers plenty of room for a *lot* of possibilities for the origin of alien invaders.

Hollywood passionately loves a good alien invasion tale; the popularity of the 1970s television show "The Invaders" and of the 1996 movie *Independence Day* are prime examples of fascinating fictional invaders coming from somewhere in interstellar space. In both of those productions, the aliens are up to no good and death and mayhem are enthusiastically presented. A bit more restrained, but perhaps *because* of that restraint even more horrifying, is the famous short story "To Serve Man" by Damon Knight (1922–2002), which appeared in the November 1950 issue of *Galaxy Science Fiction Magazine*. There the aliens appear, at first, to have the welfare of Earth people at heart as, after all, their written manifesto is a book titled *To Serve Man*. At the end of the tale, however, we learn the manifesto is actually a cookbook!

While today's science fiction aliens can come from anywhere in an enormous interstellar universe, back in the pre-twentieth century the scientific understanding of the size of the universe was as something *much* smaller. Our *lone* galaxy, the Milky Way, *was* the "Universe," and since the idea of alien invaders coming from a very distant place was in conflict with the then only imagined means of space travel—shooting a projectile out of a cannon à la Jules Verne's wonderful 1865 novel *From the Earth to the Moon*—the aliens had to come from within the solar system.[2] To the Victorian mind, then, the obvious origin of invaders had to be a "near-by" planet like Mars (which, at its closest approach to Earth is "only" a bit more than 35 million miles away, a distance it takes light *3 min* to travel) and, as is well-known, that was the idea used by H. G. Wells in his 1898 classic *The War of the Worlds*.

I must tell you, before going further, that Mars was *not* the origin of invaders in the earlier 1892 novella *The Germ Growers*, by the Australian clergyman Robert Potter (1831–1908). In that almost impossible to read theological

[2] Verne's novel is packed with fascinating mathematical and physical ideas, and a good way to fully appreciate those ideas is to read the novel as it is presented by Walter James Miller in his *The Annotated Jules Verne*, Gramercy Books 1995. The mathematical physics of *rocket* flight, as opposed to the ballistic science of artillery cannons, wasn't developed until long after Verne wrote, with the work of the Russian analyst Konstantin Tsiolkovsky (1857–1935). The math and physics of alien spacecraft equipped with the science fiction (so far) faster-than-light hyperdrives needed to travel the enormous distances of interstellar space still needs to be worked out—but theoretical physicists are hard at work on that! See, for example, Section 6.5 ("Cutting and Warping Spacetime") in my book *Time Machine Tales*, Springer 2017, pp. 326–333.

mish-mash, the aliens (who are attempting to kill all humans by spreading disease) "are inhabitants ... of [the] ether." The *ether* was imagined by the Victorians to be a mysterious substance that existed everywhere and through which the stars, planets, and comets moved with zero resistance. Its sole purpose was as something in which light could travel through in what otherwise appeared to be empty space. The ether was, in fact, what Oliver Lodge was trying to detect with his whirling machine experiments (see note 5 in Chap. 1). Modern science long ago discarded the ether as non-existent.

When Wells read Potter's work, he must have been simultaneously amused and appalled at Potter's utter lack of story-telling ability. Still, Potter's failure may well have been the spark that set Wells to thinking about how to properly do an "invasion from another world" tale. And then, 2 years later in 1894, he found his inspiration. Wells was not just a wordsmith, but also a student of science who had received a degree in zoology from the University of London. Even after becoming a writer of short stories and novels, Wells kept up with the world of science with a regular reading of the British journal *Nature*, and on page 319 of the August 2, 1894 issue, he read a short editorial note titled "A Strange Light on Mars."[3] There was mention in that note of the possibility of Martians attempting to signal Earth via the light, but Wells had an even more fantastic idea: what if the light was, instead, a "blaze [that] may have [signaled] the casting of a huge gun, in the vast pit sunk into their planet, from which their shots were fired at us."?

The "shots" were, of course, ballistic projectiles containing the Martian invaders who, once on Earth, quickly proceeded to deliver massive murder and destruction on anybody or anything in their path.[4] Wells finally puts an end to that by flipping Potter's idea upside-down: Wells' invaders are killed not by weapons of war but by common *human* germs. As a tribute to Wells' great story-telling ability, even though we now know there are no Martians *The War of the Worlds* is still a great read, and over a century later (with Steven Spielberg's 2005 movie of the invasion) it *still* scared the hell out of a lot of people.

So, here is a *double* challenge for anybody wishing to follow in Wells' footsteps: (1) Can one write yet another alien invasion tale and (2) Could that tale be free of monstrous gore? That is, does the author of an invasion story *really* have to kill millions, vaporize all the major cities of the world, melt the ice

---

[3] Wells' narrator in *The War of the Worlds* provides a citation to the *Nature* note that tells readers everything but the page number.

[4] The narrator of *The War of the Worlds* tells us that the Martians "calculated their [journey to Earth] with amazing subtlety—*their mathematical learning is evidently far in excess of ours* [my emphasis]." So here we see Wells using skill in mathematics as a way to signify superiority.

caps, and (if need be) disintegrate the moon? Could an alien invasion story instead be just a bit more up-beat? (That last line may strike you as odd, but bear with me.) Thinking about such questions soon led me to write "A Quiet, Rainy Afternoon," which appeared in the July 1979 issue of *Analog Science Fiction/Science Fact Magazine*.

"A Quiet, Rainy Afternoon"

Little Jackie Hawkins was one heck of a bright kid. He was also, like most small boys, a terror for his mother on rainy days. Trapped in the house with her son, Mrs. Hawkins' feelings for her offspring evolved from maternal love at 8 am, to mild irritation by 10, to a fairly intense unhappiness as both hands on the clock pointed skyward.

"Jackie, *get out of the kitchen*! You've been in there eating every ten minutes, you'll spoil your appetite for dinner, and I'm not going to watch you waste good food tonight. You hear me, Jackie?"

"Ah, ma, there's nothin' else to do! Pete's gone with his folks to visit his aunt, Doug's got a cold, and Tom's mother won't let'm play till his ankle gets better."

"What's wrong with Tom?" Mrs. Hawkins was feeling frustrated. *The neighborhood kids all have other things to do when the weather is rotten, but on nice days when I'd like to have the house to myself, the whole darn bunch of them clamor from morning to night to come in and play with Jackie. Where are they when you need them!?*

"He sprained it when he fell out of a tree last week. He's really got a neat swelling on it!"

"Yes, I'll just bet he does," replied his mother, "and he's probably lucky he didn't break his silly neck. I ever catch *you* in a tree, Mr. Jackie Hawkins, and—"

"Yeah, I know, ma, I know. You *tell* me all the time!" Jackie came into the living room where his mother sat, an open book in her lap. With a mouth full of cookie fragments, his rhythmically churning jaws reminded Mrs. Hawkins of a cement mixer, and the noises emitted by her son didn't do anything to hurt the analogy,

"Don't you *dare* get crumbs on the rug—I just vacuumed in here!"

"Ah, ma, there's nothin' *to do*!" Jackie flopped down onto the sofa, and stared sullenly out the window. Rain fell with the fury only a good New England spring storm can have, and there was a high probability, as Jackie well knew, that it would rain the rest of the day, and all night, too.

"What about that new friend of yours, Raymond, isn't that his name?"

Jackie made a face. "Ray and me had a fight. We ain't friends anymore."

"Ray and *I*, Jackie, not *me*. And don't use *ain't*. Your grammar is terrible. Now say it right."

"Ah, ma—Ray and *I*. I'm goin' downstairs and watch TV!" With that, Jackie slouched off the sofa and, as his mother watched with relief, he disappeared down the stairs into the basement game and TV room.

Jackie didn't really want to watch TV, but things *were* pretty dull. He flipped it on, plopped down on the floor, and gazed unhappily at the screen. It was just dumb old cartoons! He started to reach for the channel selector when the horizontal sync generator took that instant as the best time to blow its 15.75 kHz oscillator. The unsynchronized scan lines made a blotchy colored mess on the 25 inch display.[5]

Any other boy on Jackie's block, or in the whole town, for that matter, would have just turned the set off in disgust. But not Jackie. He was a heck of a bright kid, and not one to be intimidated by a lack of specialized knowledge. As he watched the kaleidoscopic pattern of dazzling patterns, he knew what he could do to pass the boring day. He'd fix the color TV.

He craftily noted that as long as he was reasonably quiet, and responded instantly to any inquiry from his mother, she'd probably not investigate. That'd give him almost 4 h until his dad came home from work. Why, that'd be *plenty* of time to fix something simple like a color television set! He turned the power off, and pulled the set away from the wall. He was ready to start.

Intelligent lad that he was, Jackie immediately pulled the power plug out of the wall. He knew better than to fool around with the power still capable of zapping him. Then he tip-toed up the stairs, and cautiously peeked around the corner into the living room. Mrs. Hawkins was deeply engrossed in her book, and her reading glasses were already slipping down her nose. Jackie knew his mother would soon be asleep. It was perfect.

Quietly, very quietly, he turned back down the hallway, to the combination closet/tool area his father kept. Jackie looked briefly at the neat, well-stocked shelves, and then selected two screwdrivers (flathead and Phillips), wire-cutters, needle-nose pliers, a roll of rosin-core solder, and the 125 W soldering gun. His arms cradling the loot carefully, he again tip-toed his way back downstairs to his electronic patient, ready to begin surgery.

Removing the plastic back cover, he gazed at the maze of solid-state electronics, wire-cables, and complicated mechanical linkages from the tuner

[5] These words somewhat date the story: while a horizontal scan-line sync generator commonly appeared in the cathode-ray tube sets of the pre-2000s, modern TVs with solid-state displays don't need it. Nevertheless, I decided to leave it in here for "historical integrity." Similarly for the "mechanical linkages" mentioned a bit further-on (which will mean nothing to readers used to electronic remotes).

switch to the AM video and FM audio circuitry. It bothered him not a bit. He'd heard his father say, many times, when confronted with balky electronic appliances, "Ninety-nine times out of a hundred, Jackie, the problem is caused by a bad connection. Just check all the contacts, clean 'em up, and that usually does it!"

Jackie pulled all the integrated circuit chips out of their sockets, and very carefully scraped each pin with a screwdriver blade. Then he put them back—all in the wrong sockets. It wasn't hard to do—after all, a three-stage Butterworth filter chip *does* look a heck of a lot like one with ten cascaded operational amplifiers on it.

Then Jackie plugged the soldering gun in, and reheated all the exposed joints on the tuner, just to be sure there weren't any cold joints. He didn't know what a cold joint was, or even comprehend the nature of a broken circuit, but he'd heard his dad talk about those, too. So he touched the gun to every joint he saw, adding generous globs of fresh solder for good measure. And the molten metal on at least ten of the joints flowed enough to short-out with each other. Some *very* peculiar circuits were thus created.

Finally, just to be sure there were no problems with the antenna, he unscrewed connectors to the 300-ohm lead-in wires and polished their contacts. Then he put them back—with one half of the antenna on the correct terminal and the other connector to the normally unused UHF 75-ohm terminal. This set-up the electrical geometry for some mighty curious standing wave patterns in the antenna circuits, but, of course, Jackie wasn't the least bit concerned. He hadn't even *heard* of a standing wave.

Satisfied with his work, Jackie screwed the back cover on, plugged the power cord in, pushed the set back against the wall, and flipped the power switch on. It had taken him some time for all these corrective actions, and his father would be home soon. He had to hurry.

It's too bad the top physicists and electrical engineers in the country weren't there. Then they would have seen the first Mosaki Sub-Space Quantum Modulator in operation, 232 years before Philip Mosaki rediscovered it. Yes, *rediscovered*, because by an incredibly amazing series of accidents, that's what Jackie Hawkins had turned the color television into. And a good thing he had, too.

For just at that instant, the Vodarian battle cruiser *NEBUTRON* came out of faster-than-light hyperspace, shifted over from its Mirchlen neutrino drive to the sub-light photon units, and went into a polar orbit 72,000 miles above the Earth. Sent by the Vodarian Space Council to observe the activity of all (if any) intelligent life in advance preparation for invasion, the Vodarians were met by a startling surprise.

"Supreme Commander, the sub-space sensors are recording an unusually high quantum modulation level! Previous robotic surveillance of this star system indicated no such technologically sophisticated life forms present. The master computer must have a data recording error."

"Hmmm," burbled the Supreme Commander, "to proceed with our pre-invasion orders, in view of this new knowledge, *could* be dangerous. What do you suggest, Squamus?" The Supreme Commander's aid was, as are all Vodarians, very cautious.

"I suggest, Supreme Commander, that we tentatively probe their quantum modulation field. If we can cause an appreciable breaking of the phase-lock amplitude, then we are more advanced. If, on the other tentacle, their phase-lock *cannot* be broken, we should withdraw immediately!"

The Supreme Commander had won seven battle stars by following the advice of his aid. No reason now to break the chain of luck. "Proceed, Squamus."

Turning to the battle cruiser's control center, the aid projected the orders by mind thought—his wishes were carried out immediately by the *NEBUTRON*. But young Jackie Hawkins had done his work well. The phase-lock tracking sub-system he had created in the TV set modulator would have swelled Philip Mosaki's heart with pride—if he had been there. But, of course, he wasn't, since his great-great-grandparents hadn't been born yet! Even as the Vodarian sub-space prober probed and prodded, Jackie's modulator counter-probed and counter-prodded with hardly a miss of a hertz in frequency.

Back on Earth, Jackie was perplexed. The screen was *really* messed-up now, and there were strange burbling sounds coming out of the speaker. Boy, was dad going to be mad! For the first time, Jackie began having self-doubts. He was, after all, just a small boy. Scared of what his father was going to say, Jackie scooped up the tools and ran up-stairs to put everything back in the closet, all neat and proper. But he wasn't fast enough—his dad came though the breezeway door with a booming "Hi, everybody, I'm home. Boy, is it dumping out there!" Jackie rushed into the kitchen to keep his father away from the basement stairs.

"Hey, Jackie, how's my boy? Do anything interesting today?"

"No, nothin' really, dad. Just took it sorta easy." Jackie prayed his father would stay out of the basement until he could get back down there. He had left the television on, with the screen all crazy and making strange, really weird noises.

"Where's your mother, son?"

"In the living room. Say, dad, let's go in and see ma. I'll bet she'll be real glad to see you!"

Mr. Hawkins looked startled at this suggestion—Jackie was at the age where little boys didn't usually desire to have conversations with parents, with the exception of a short demand, once a week, for their allowance. He walked to the entrance of the living room, and saw his wife sound asleep, her book on the floor and her glasses in her lap.

"No, let's let your mother sleep, Jackie. She's had a tiring day, too, I'm sure." Mr. Hawkins was a thoughtful man. "Come on, son, let's go downstairs and play a game of ping-pong. I'll spot you five points!"

"But dad—" and then Jackie resigned himself to the inevitable. His father was gone, disappeared down the stairs. Jackie glumly started down to meet his fate.

"Oh, my—oh my God! What the heck is *this*!?" Jackie didn't need to guess what his father was yelling about. But as he came into the room behind Mr. Hawkins, his eyes widened in amazement, too.

The television set was no longer merely flashing pretty colors and making burbling sounds. It was glowing with a dull red-purple halo, while seemingly fading in and out in substance. First Jackie *thought* he could see the back wall of the room through the set, and then the set would become solid again. And then it would *appear* to fade once more, the cycle repeating every 5 s or so. What had happened, as would be theoretically derived by Philip Mosaki in a couple of hundred plus years, was that the third and fifth harmonics of the sub-space quantum field were resonating in a coupled energy-exchange mode. A most dangerous mode, too, one that would serve as the basis for the deadly Mosaki ripping beam, standard armament on all Earth starships of the late twenty-third century.

So it was, therefore, that even as Jackie and his dad watched in horror and fascination at the strange antics of their television set, the *NEBUTRON* found itself in a battle for survival. "Supreme Commander," burbled Squamus in fear, "the aliens are attacking! Their matter disruption beam is incredibly intense, the ship's defense shields are being inexorably torn to shreds. We *must* retreat!"

"Hmmm," burbled the Supreme Commander but, before he could continue, the left photon thruster tube of the *NEBUTRO*N was yanked off and dissolved into a rapidly expanding mist of vaporized metal. That was enough to decide the issue. "Squamus, GET OUR ZLOBS OUTA HERE!"

But it wasn't to be that easy. The resonating fields of gigantic quantum gradients were now quickly dismembering the *NEBUTRON*. The battle cruiser just made it back into hyperspace before the last Mirchlen neutrino drive generator fused, but not before it was severely damaged. All the way back to Vodar the *NEBUTRON* flicked in-and-out of hyperspace for brief moments, each time losing huge quantities of hull material in collisions with cosmic dust. By the time they reached Vodar, the hull was only one-tenth its original thickness and the *NEBUTRON* had a noticeable sag in its profile. In addition, her crew was sterile. Hit with the hard radiation generated by the hull interactions with ordinary space at hyperspace velocity, they would produce no new little Vodarians.

The Space Council, naturally, had no recourse but to haul the Supreme Commander up on a general court martial for bringing a battle cruiser back in shabby condition, with an infertile crew. There was no possible defense. So, they busted his zlob right out of the space service. And they canceled the invasion of Earth. Vodarians *are* cautious.

Meanwhile, moments after the *NEBUTRON* first vanished into hyperspace, the Hawkins' television set overloaded and blew every breaker in the service entry box to the house.

The internal electronics of the set, so recently converted into a Mosaki Sub-Space Quantum Modulator, melted. As the future Philip Mosaki himself would discover, such an awesome physical phenomenon needs proper cooling, something Jackie hadn't provided. But, after all, he was a very small boy and didn't know any better.

"Well, I'll be blasted! I never saw anything like that! I'm going right down to Sears[6] tomorrow and get a replacement—good thing the one-year warranty is still in effect." Jackie's father looked puzzled but, to Jackie's profound gratitude, not a single suspicious glance came his way.

Later that night, just before he drifted off to sleep, Jackie thought with eager anticipation about the next day's weather forecast—received over the kitchen radio since the television was, of course, incapacitated. No rain! Boy, tomorrow would be fun, not dull like today. On rainy days, there's just *nothin'* to do!

In this story, the concept of *hyperspace* appears once again. See note 3 in Chap. 1, where I mention that the idea of hyperspace was hinted at by Wells as early as 1895, long before science fiction realized hyperspace was *necessary* to sidestep Einstein's limit on speed as that of light. In "The Remarkable Case of Davidson's Eyes," a man can see through a "kink in space" to the other side of the planet. Wells has a character in the tale suggest this is due to "some extraordinary twist given to [the man's] retinal elements [caused by] the sudden change in the field of force [caused by a lightning bolt while] stooping between the poles of a big electromagnet." When presented with this

[6] When I wrote this story, Sears was still in business. The store did declare bankruptcy in 2018 but. Since it doesn't appear (yet) to be *totally* dead, I've left the Sears mention in without change.

explanation, the narrator of the story says "To talk of there being 'a kink in space' seems mere nonsense to me; it may be because I am no mathematician." When the narrator says "Nothing would alter the fact that the [man was seeing things] eight thousand miles away," the reply was "[T]wo points might be a yard away on a sheet of paper, and yet be brought together by bending the paper round."

Figure 3.1 shows what Wells had in mind, with **A** and **B** being two points in a 2D space that is *curved* in a 3D "hyperspace." Wells' narrator rejects this, saying "The reader may grasp [this] argument, but I certainly do not." Others have, however, and science fiction has enthusiastically embraced it!

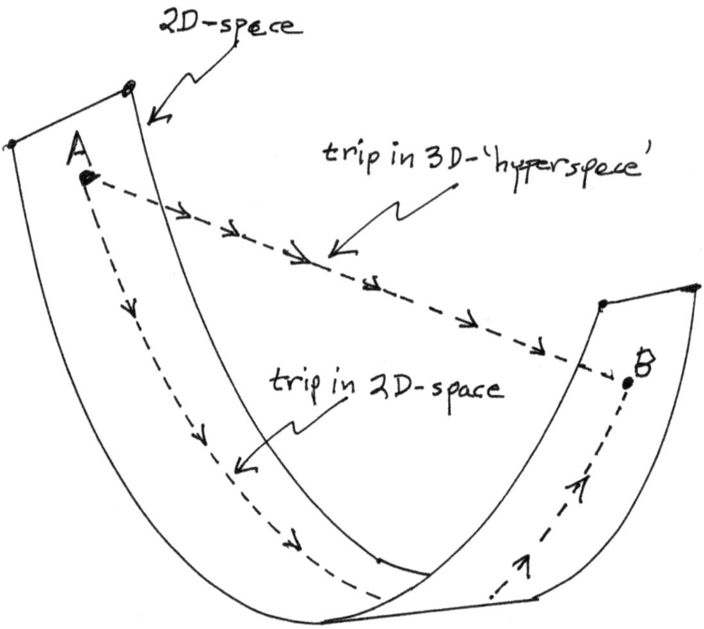

Fig. 3.1 A trip from **A** to **B** that is confined to "normal" 2D space is *longer* than is a trip that leaves the 2D space at **A** and travels through 3D *hyperspace* to **B**. The trick to make this work, of course, requires learning how to bend 2D space!

4

Revenge

Among the most powerful of human emotions is the desire for revenge, for an opportunity to deliver a "payback" in response to some offensive event suffered in the past at the hands of a transgressor. Maybe it was something said or done just yesterday, but the hatred might also be in response to a foul deed experienced decades ago. Revenge motivated by events lying years in the past may be the most powerful of all, in fact, as those events have had a very long time to fester and to corrupt the soul of the injured party. Revenge-delayed may also result in the most violent of paybacks, in full agreement with the well-known saying that "revenge is a dish best-served cold." Such "long-time-coming" paybacks may be the most satisfying of all.

In a non-science fiction setting, the form of a violent payback may be one of the several obvious possibilities: stick a bomb in the target's car or under his bed, get a gun and shoot him, drop a packet of rat poison in his dinner, toss a plugged-in electric fan into the tub when he is taking a bath, and so on. You see that sort of thing on television, in the movies, and in novels all the time. You even see such things, now-and-then, on the evening news.

A *science fiction* revenge, however, offers up exciting new possibilities! In the early days of science fiction, for example, the availability of time machines was often too seductive to pass-up. The common approach was for the aggrieved to use the machine to travel back in time and then, after locating the target and doing away with him/her, returning to the moment just after the moment the time trip started. There might be some extra story frills here and there, but that was the core of such tales.

An example is "The Men Who Murdered Mohammed" by Alfred Bester (1913–1987), which appeared in the October 1958 issue of *The Magazine of*

© The Author(s), under exclusive license to Springer Nature Switzerland AG 2026

P. J. Nahin, *The Art of the "What-If?"*, https://doi.org/10.1007/978-3-032-15618-1_4

Fantasy and Science Fiction. In this tale, a professor comes home from school to find his wife in a passionate embrace with another man. This so enrages him that he (after inventing a time machine in his home office in just $7\frac{1}{2}$ min) travels back in time with a gun to murder his wife's grandparents before their children will be born. He is sure that will prevent his wife from having been born, but this turns out not to be so. Indeed, before the story ends Marco Polo, George Washington, Napoleon and, of course, Mohammed, are also killed, along with "half a dozen other [historical] celebrities," and yet the professor's wife remains in her adulterous embrace. The past has most definitely been changed, but the present is unaltered.

Bester offers up a convoluted word salad "explanation" for this (involving another time traveler), but none of it is very convincing (that is just my opinion, of course—Bester's tale is, in the science fiction community, generally held in high regard). The story, however, comes across to me mostly as a joke: the professor is on the faculty of the Department of Applied Compulsion at Unknown University, which is almost as big a clue to the tongue-in-cheek spirit of the story as is the rapid construction of a time machine from the odds-and-ends lying about a home office. Despite all the murders, it really isn't a *dark* story, something I think a serious revenge tale has to be.

To write such a time travel revenge story today just would not work because it is now widely recognized (by both philosophers and physicists) that self-consistent physics means the past *can't* be changed. Of course, if history has recorded that the target *had* been killed long ago by an unknown assailant, such a tale *could* work, as all the trip back in time would do is establish that the once mysterious killer was the time traveler. That does leave open the question, however, of what is the purpose of even making the trip, as the target is *already* dead before the trip begins!?[1] A pursuit of this and related questions quickly gets us into discussions of the paradoxes of time travel to the past, which I'll defer until a later chapter.

When I first started to think about writing a science fiction revenge story, I quickly decided that using a time machine was just too much in the category of "enough already." I wondered if, instead, a science fiction revenge story

[1] Just as a hint of the sort of mind-twisting situations that time travel to the past drops us into, this puzzle is an illustration of the distinction between *changing* the past (not allowed) and *affecting* the past (allowed). The vengeful time traveler *has* to make the trip back in time because he *did* make the trip (because that is why his target is dead *before* he starts the trip). This, however, now prompts the question of: what if the time traveler decides *not* to make the trip? If you believe in free-will, he certainly *can* make that decision. But, for self-consistent physics to hold, there must be *something* that *compels* him to *not* make the decision to not make the trip. Oh, my, that is too many *nots*—my head is exploding! Think about this while we work our way toward the time travel chapter.

could be done without the use of *any* super-science gadget at all. It didn't take long before I realized that the answer is *yes*, because Arthur Porges (see note 2 in Chap. 1) had already done that (twice!) with his tales "Off His Rocker" (in the February 1960 issue of *Fantastic*) and "Revenge" (in the February 1961 issue of *Amazing Stories*). Both tales have an unhinged scientist responding to what he feels was ill treatment by powerful, impersonal corporate forces. One of them uses his knowledge of biochemistry to achieve his goal, and the other is a physicist who uses his understanding of mechanical resonance to heap disaster on his target.

Porges' tales are good reading, but both lack the very personal, one-on-one desire for revenge that I wanted to power *my* story. In "Revenge," for example, the narrator (the scientist) specifically worries that, as a result of his revenge, he risks being considered a mass murderer. In the story I wanted to write there would be *no* such concerns about "going too far." I wanted to write a *very* dark story, indeed.

The result was "The Reunion," which appeared in the April 1979 issue of *Analog Science Fiction*. The title refers to events that occur countless times, every year: the reunions of high school classes. And where else but in *high school* do events occur, on a daily basis, that are mean, hurtful, and often downright cruel and sometimes even evil? Events guaranteed to fester and boil with increasing heat as the years pass. Events that, with near virtual certainty, will lead to a perhaps overwhelming urge to seek deadly revenge.[2]

Most of the time, of course, such urges remain just a fantasy. But, now-and-then, they become something more; possibly much more, as in "The Reunion."

"The Reunion"

Dr. Richard C. Breed walked briskly across the campus parking lot to his car. He casually greeted two of his women students, and glowed inwardly at their open admiration of his carefully nurtured, sophisticated, professorial image. Teaching the college service course in geology certainly had its benefits. He *did* cut quite a figure, dressed in 55-dollar[3] wing-tip shoes, immaculately tailored vested tweeds, and corduroy hat with a jaunty red feather in the brim. The silk tie, silver cuff links, and 500-dollar pocket watch with gold chain (a

---

[2] The gold standard for high school revenge stories *has* to be, of course, Stephen King's 1974 fantasy horror-thriller *Carrie*. The supernatural aspect of that work was, however, something I couldn't make use of in what I wanted to be a *science fiction* tale.

[3] This story was written decades ago; today, all the monetary numbers in it would probably have to be tripled (at least) to still be realistic, but I have left them as originally published.

Phi-Beta-Kappa key hanging prominently from it) completed the picture. He puffed with pleasure on his 200-dollar Danish briar pipe, leaving the fragrant, faintly vanilla flavor of "Captain Black" in his wake. The women loved it, and with his rugged red beard, he liked to imagine himself a scholarly Viking.

Richard Breed had it made. A scientific entrepreneur who regularly had his $100,000 a year NSF geological research grant renewed. He didn't do much original field work himself now, of course, busy as he was recruiting new doctoral students to do it for him, and hustling foundation money to pay them. Oh, once in a while he might chip rocks at a local site, but reading seismological activity charts in his air-conditioned office fitted his schedule better. Money was definitely not a problem for Breed, either, not with the additional summer salary of two-ninths of his regular (already very substantial) academic-year pay!

Yes sir, he had come a *long* way from the fat, pimply teenager with bad breath he had been in his small, Southern California high school of almost a quarter century ago. He frowned as he thought of those long-ago schoolmates that he had once envied, and hated. Young, and arrogant with their supposed immortality, they had never let him belong. He had been the outcast, the one made to feel like something peculiar, almost repugnant. While the others had greeted each other loudly in the halls, gone to dances together, and generally gloried in the golden days of teenage youth, he had spent those same years as a lonely recluse in the library, alone with his science and math books. Called "four-eyes" and "specs" until the words made him ill, he had finally started to come out of his shell late in his senior year, when he won a national math contest. But it was too late to make any real difference then.

Even now, as a 41-year-old late bloomer, who had certainly bloomed well, his physiological reaction to recalling those 4 years of pain was a blush of shame and embarrassment at the klutz he had been. And an almost overwhelming flood of hate at those who had made him so miserable. "Ah, if only those bastards could see me now. The ones that didn't become mothers at eighteen, or criminals, are probably eight-to-five office slaves or beer-paunchy sanitation workers. If only I could see *them* now, it would be so very, very different!" The thought ran sweetly through his mind, and he enjoyed the imagined pleasure of what he knew would be their envious reaction to the new, soon to be famous Richard Breed.

He reached his car and carefully placed his bulging briefcase on the back seat. Stuffed with lecture notes and color slides for next week's International Seismological Conference in Montreal, he planned to study them one more time over the weekend. He was absolutely sure of his calculations, but what he was about to reveal was such a shocker he knew the questions would come hard and fast. He had to be overpowering with his rebuttals.

After Montreal, election to the National Academy was certain. The Nobel Prize in physics would surely soon after be his. With a change in Administration, he could even imagine the post of Presidential Science Advisor being offered. And maybe he'd take it, and maybe he wouldn't!

Once home, after checking the mailbox, he locked the conference papers in his desk to await a careful restudy later that evening. Tossing the handful of that day's letters on his desk, he went into the kitchen and mixed himself a Bloody Mary. As he sipped his drink while walking back to the study, he let his thoughts run on over the pandemonium he knew would result at the bombshell he would soon set off in Montreal.

After years of painstaking data gathering and several extensions of Carson's dissertation ("I get the Nobel and all he gets is a PhD and an assistant professorship somewhere, but that's the name of the game!"), Breed had formulated a mathematical prediction theory for earthquakes. His test calculations over the last 3 years had all been on the mark in location, time, and intensity. But telling people you had privately predicted quakes that had already happened wouldn't cut any water. He had to put his butt on the line, in public, with a major *prediction*.

The new strain-gauge data from Caltech's monitoring stations along the San Andreas fault had given him what he needed. On the morning of June 14 of next year, there was going to be a monster 8.3 shake with its epicenter 40 miles south of Los Angeles. Breed chuckled to himself as he thought how close that was to his old high school. Mother Nature was going to do what he had wanted to, so many years ago.

What was really going to cause trouble wasn't just that damn near all of Orange County was going to be flattened. If they didn't listen to him, it was also going to explode! Right in the middle of the target area, next to the sleepy little town of his youth, was a vast underground storage facility of liquified natural gas. Used as a depository by several of the oil giants, the town was literally floating on a lake of LNG. The entire county would have to be evacuated, and the tanks drained. The cost would be horrendous. Breed figured they'd evacuate people, all right, but would take a chance on leaving the tanks full. What a show that would be! Weather satellite imagery of the detonations would be the lead evening news stories for days, worldwide.

He began to thumb through the mail on the desk as he finished his drink. All the usual crap was there—electric bill, a reminder to stop at the bank and have his credit card photo updated, and his subscription copy of *Time*, only 3 days after it had already hit the newsstands. Then his eyes caught a familiar return address on a stapled flyer and he held his breath as he unfolded it.

> *MEMBERS OF THE CLASS OF 1957*
> *Thanks for your great response to our previous announcement of the plans for our TWENTY-FIFTH REUNION next year! Can you believe it's been almost a quarter-century since we last walked the halls of high school together? We can't either! But we still haven't heard from everyone! Out of 93 announcements, we got 82 replies. If you are one of the missing eleven listed below, or know how to reach someone on the list, please let us know. The reunion dinner and dance will be the evening of June 13, and the next morning, too! The place, of course, will be the high school gym. Singles are $12, couples $20, for dinner, drinks, and dancing.*
>
> *Looking forward to seeing you*
>
> *CLASS OF '57 REUNION COMMITTEE*
>
> *David Sliply*
>
> *Tim White*
>
> *Mary (Stone) White*

Breed was stunned for a moment, as he saw his name among the missing 11. Christ, he *would* be able to put those SOBs in their place! But what was all this business about a prior announcement? He hadn't seen it or heard anything. Ah!, now he knew what must have happened. For 2 months earlier in the year he had been abroad. He had instructed his secretary to forward only the obviously important stuff, and to hold everything else. The first announcement must have gotten lost in the pile of local junk mail and campus events notices.

Breed's eyes wandered over the sheet again, and stopped at the names of the reunion committee. Wouldn't you know it—the three biggest creeps in a class of creeps. Memories came back in a wave of pain:

It was a cold, windy day, and the third period gym class had just suited up. The boys were on one side of the football field, and the girls on the other, all lined-up for roll-call. As usual, fatso Dick Breed felt awkward and out of place in gym shorts (number 64) and T-shirt. A flaming red-head, he didn't tan, and his skin was pasty white. The green shorts looked like a bacteria mold growing on his bread dough body. He had been late to class, and as he stripped and dressed hurriedly, while thinking about next period's trig exam, Breed had failed to notice his shorts were on backward.

But that primitive Neanderthal, Dave Sliply, noticed. "Hey, Four-Eyes! Does that 64 across your butt stand for its weight or its size?" The girls heard it, too, and even the gym teachers grinned. The entire class laughed, and

Breed could have died with shame. The rest of the period passed in a horrible, slow crawl, all the while that damn number burned itself into his backside.

The senior Prom was all that everyone had been talking about the last 3 weeks. Dick Breed had been walking around with butterflies in his stomach for days, especially in English IV, where he sat behind Mary Stone. Time and again, he had almost worked up enough courage to ask her to go with him, and then something happened to ruin the moment. But today he promised himself he would ask. Time was running out. Mary came into the classroom with a group of giggling friends, and Breed braced himself. As she sat down in front of him, and before she could start talking to anyone else, he leaned forward and spoke in a not-quite-a-whisper voice. "Uh, Mary, I was, uh, sort of wondering, that is if you don't have anything else planned, if you'd like to go to the Prom dance? With me, I mean!"

For a terrible, sick moment, Breed thought she was going to just ignore him. Later, he wished she had. Mary turned slowly in her chair until she almost, but not quite, faced him. "Dick Breed, you little frog! How *dare* you ask me that, right here in front of my friends!" Mary was hissing the words out in a tight, hushed voice. Her face was distorted with rage, and Breed was actually scared. The students around them began to notice the exchange, and Breed tried to shrink through the seat of his chair. "I'm going to the Prom with Tim White, and if he found out about this, he'd take care of you, but good! So you just keep away from me!"

That night, for the first time since he was a little boy, Breed cried himself to sleep.

Breed was sitting alone in a corner of the cafeteria, eating that day's cruddy, mass-produced meal of greasy taco, reburned beans, salad turned brown at the edges, and warm lemonade. He scarfed it up with the indiscriminating appetite of the teenage overeater. He could hear all the other kids at the nearby tables yelling in good-natured cheer. Nobody talked to Breed. Suddenly he heard his name called.

"Hey Dick—Dick, over here!" Breed looked up from his tray in surprise, and saw Tim White waving an arm at him. Big, tall, handsome Tim White, who had taken Mary to the Prom. "Hey Dick, Jack Preston tells me you're pretty good with your hands in machine shop. Is that true?"

Breed couldn't believe his ears. Maybe *this* was the key to getting to be friends! He'd helped Preston chuck a piece of tubing in a lathe last week, in second period shop, and had shown him how to cut threads in one end. The word must be getting around that fatso Breed wasn't a total creep.

"Uh, yes, I guess I'm okay, working with tools. You know I'm not great or anything like that, but I do all right, I suppose." He saw a little smile on Tim's mouth, and that worried Breed. Was this just another damn trick?

"Well, I'll tell you Dick, I was wondering if you wanted to come with us on a deep-sea fishing trip next weekend. My dad's renting a boat at Newport and we sure could use a good man with his hands, like you." White spread his arms wide to indicate the group of grinning boys and girls sitting at his table as the fishing party. Breed's suspicions evaporated. Tim *was* just being friendly!

"Gosh, that sounds like fun, Tim. Is there anything special I could do to help out on the trip?"

"You bet, Dick! A guy like you, really good with his hands—well, while we got the fishing lines ready with hooks for the bait, you could be the one doing the baiting: *you'd be our Master Baiter!*" White collapsed in hysterical laughter, and his gaggle of hoodlum friends hooted at Breed who, burning with shame, ran from the room.

Professor Richard C. Breed, PhD, sat motionless for a long time, his right hand wrapped around the Bloody Mary in a sweaty grip. Being a sturdy tumbler made to hold a stiff triple shot of whisky was all that kept it from snapping. Finally, he picked up the telephone and dialed. He could hear the long-distance switch gear rattle on the circuits like electronic crickets.[4] Then the sharp, clear ringing of a telephone thousands of miles away.

"Hi, Pete, this is Dick Breed. Yeah, I'm fine, but look, I've got some late-breaking plans. I'm not going to be able to make it to Montreal. I've just turned up some problems in my data. No, no problems in the theory. Just some technical details in my computer runs, but I would like to have it just right before I present it publicly. As conference chairman, I hope you can fill in my scheduled slot and give my regrets to the others. Okay? Hey, great Pete. I appreciate it. I'll try to get my stuff cleaned up for the go around in England next year. Yeah, I'd like to spend the Christmas holidays in London, and that'll be a good reason. Thanks for covering for me, Pete. My best to Betty. So long."

Breed hung up the receiver, stared at the reunion notice, and then crumpled it in his hand and flung it into the trash bucket. "Eat, drink, and be merry, you bastards. I'll be watching for you on the evening news."

Was Breed a mass murderer? It seems doubtful (to me, in any case) that he could be legally charged in a court of law. But what of his *moral* standing? As I wrote this story, I couldn't help but be reminded of another old saying about revenge: "Before you start on a journey of revenge, dig two graves." The year after "The Reunion" appeared in *Analog*, it was reprinted in the book

[4] Ah, yes, the *dial* phones and *relay* telephone exchanges of the 1970s. Where have they all gone?

anthology *Microcosmic Tales*, Taplinger 1980. One of the editors, Isaac Asimov, wrote some lead-in words for each of the 100 stories, and the ones he wrote for "The Reunion" were, alas, as close as I ever got to "meeting" Asimov.

As a final comment on revenge, fictional story-telling itself can be a highly effective method for delivering revenge on the *memory* of a target. My favorite example of this is the following fantasy treatment[5] of Hitler's fate, just after he blows his brains out as the war in Europe comes to an end in 1945. After killing himself, Hitler instantly goes to Hell (no surprise there), with the Devil himself waiting to greet him, in a hallway lined with three doors. The Devil tells Hitler that, despite what most people believe, Hell is a "civilized" place, and that each new arrival is given a choice for what will be the eternal punishment they will suffer in retribution for living a monstrous life. For Hitler, there are three possibilities.

Taking Hitler to the first door, the Devil opens it and Hitler sees Hermann Göring nailed to a slowly rotating wheel that is alternately dipping the war criminal into and out of a kettle of boiling oil. The screams are horrible, and Hitler turns white.

Taking Hitler to the second door, the Devil opens it and Hitler sees Heinrich Himmler nailed to the floor, covered by a huge cloud of monstrous, stinging wasps. The screaming is even worse, and Hitler almost faints.

Taking Hitler to the third door, the Devil opens it and Hitler sees Joseph Stalin[6] making love to a beautiful woman. Hitler instantly becomes euphoric, and excitedly tells the Devil "Oh, boy, this is the one for me! I choose Stalin's punishment!"

"You sure?" asks the Devil. "Remember, this is *eternal* damnation."

"Yes, yes," yells Hitler, "this is my choice! You promised, it's *my* choice!"

"Okay," replies the Devil, with a sly grin, "you've got it. But just one thing—it's actually the lady's punishment."

[5] An elaboration of a story told in the 1986 novel *Stallion Gate* by Martin Cruz Smith (describing events in Los Alamos, just before and up to the Trinity test of the first atomic bomb).

[6] When Hitler died Stalin was still alive and so not (yet) in Hell, but that is just a minor quibble in a fantasy tale.

5

Invisibility

Second only to time travel, the most thought-provoking story theme in science fiction is, I believe, that of the "invisible man." Both themes are strongly associated with H. G. Wells, who wrote the two classic novels that instantly come to mind when talking of time travel or invisibility. In the 1895 *The Time Machine*, Wells interprets the fourth dimension as time, but in the 1897 *The Invisible Man* the fourth dimension is taken to be spatial. In that second novel the central character "explains" the secret of his discovery this way: "I found a general principle, a formula, a geometrical expression involving four dimensions." I'll say more about both time travel and the fourth dimension in later chapters, and so here I will focus on the "invisibility" concept—and its central problem (one that Wells understood but consciously *ignored* in the novel).

Specifically, to be truly invisible the man's *eyes* must be invisible, which means they have to be completely transparent to light. But when you think about how your eyes work, how they *trap* light, that what your brain interprets as visual images (as *sight*) is the result of electrical activity produced by photons of light being *absorbed* by the retina, then the question is: how could a completely invisible man *see*?

Well, he couldn't—he would be blind!

It is clear Wells fully understood this problem. At one point in *The Invisible Man*, when the central character is describing his experiences during the moments immediately after he performed his vanishing experiment, he remembers looking through his fast-fading hands: "I could see … through them though I closed my *transparent eyelids* [my emphasis]." There is no mention of his retinas being similarly transparent, however, and that's because (I

© The Author(s), under exclusive license to Springer Nature Switzerland AG 2026
P. J. Nahin, *The Art of the "What-If?"*, https://doi.org/10.1007/978-3-032-15618-1_5

believe) Wells had no answer for it. So, taking advantage of his ultimate power as *author*, he simply *ignored* the issue![1]

Being blind is a pretty serious objection, one *so* serious, in fact, that you would think it alone to be sufficient to push the invisible man concept entirely into fantasy and totally out of science fiction. Well, maybe not. Otherwise, how to explain the tale told in *The Philadelphia Experiment*, a book published in 1979 and made into a 1984 movie of the same title? Written not as fiction, but rather as the "true" story of the US Navy's supposed experiments during the Second World War on the possibility of rendering a warship invisible (in particular, the USS Eldridge, a destroyer escort). The thesis of the book is that the Eldridge was made to vanish, in 1943, while at the Philadelphia Naval Shipyard.

It is certainly fair to say that though the movie is fun to watch, the book is 100% junk science. (So is the movie but *it*, at least, is fun to watch with a bowl of popcorn!) The general belief, today, is that the "invisibility" story started as the result of a misunderstanding of the physics of reducing the *magnetic signature* of a metal ship (the result of the ship's exposure to the Earth's magnetic field). That signature was used by the Germans to trigger the detonation of a nearby sea-mine, and the idea was to reduce the signature by properly wrapping the ship with cables carrying large electrical currents. The magnetic fields created by those currents could reduce the net magnetic signature to the point of making the ship *undetectable* by a magnetic mine.

The goal of making a ship undetectable (that is, "invisible") to a magnetic mine was garbled into making the ship *invisible*, period. Experiments in the magnetic signature reduction process (what is called *degaussing*) *did* take place during the war, and are still done today. The extension discussed in *The Philadelphia Experiment*, of achieving invisibility across the *visual* spectrum, however, is scientific nonsense. In an attempt to gain some sort of scientific plausibility, *The Philadelphia Experiment* mentions that Einstein was a paid consultant to the Navy, and makes a strong "hint" that in that work he applied his so-called unified field theory to the invisibility project. That silly assertion, alone, reveals the junk nature of *The Philadelphia Experiment*, as Einstein's work was in the general area of the physics of high-energy explosives and explosions, and certainly not in any invisibility project.[2]

---

[1] As further evidence of Wells' awareness of the "blindness" problem, see my *Time Machines* (2nd edition), Springer 1999, p. 146, for Wells' own words, in a letter written to a critic who had spotted the invisible man's vision problem.

[2] For what Einstein was *really* doing for the Navy, see Stephen Brunauer, "Einstein in the U.S. Navy," *Journal of the Washington Academy of Sciences*, September 1979, pp. 108–113. Brunauer (1903–1986), a

But for a writer of science fiction, what a great idea! Is there any way, I wondered, to use invisibility in a story in a manner that *would* make physical sense? In particular, could the "vision problem" in *The Invisible Man* be gotten around? The challenge of writing such a story initially caught my imagination when the hijacking of an airliner by a man called "D. B. Cooper" made international headlines in 1971. *How*, it was wondered, did a hijacker dressed in street clothes manage to survive bailing out of a commercial jet flying 200 mph at 10,000 feet, in a snowstorm(!)? The answer to that question still puzzles to this day. In my story "Archival Voice," which appeared in the February 1984 issue of *Analog Science Fiction Magazine*, I offered the following "possibility" for what happened.

"Archival Voice"

The Time: November 24, 1971, 7:49 P.M. [PST].

The Place: Somewhere over the Cascade Mountains of southern Washington.

The beleaguered jet flew through the swirling snowstorm, its flight crew struggling desperately to keep the battered aircraft level. Sightless due to ice an inch thick on the windshield, wipers frozen to the glass, they trusted their fate to the lights, gauges, and radar scope that provided the only illumination in the cockpit. The circular sweep of the weather radar beam showed a solid mass of intense storm energy all around the plane and so the men ignored it. They kept it turned on, though; in spite of its uselessness the steady rotation of the beam was a psychological comfort, and Lord knew they could use a bit of that. The madman in the rear of the plane would probably end up killing them all.

Suddenly the captain cursed aloud, his voice filled with helpless rage. His eyes were fixed on the steady red glow of a panel light that had just flicked on.

"Jesus Christ, the bastard's lowering the rear ramp!"

The flight attendant, locked in the cockpit only minutes before, said nothing. She was too well trained to succumb to hysterics, but the knuckles of her hands were white with the tension of her grip on the emergency jump seat. Dear God, but she didn't want to die.

The Time: March 17, 1987, 11:07 P.M. [EST].

The Place: The Blackworld floor, CIA sub-basement security vault, Langley, Virginia.

---

chemist, was the man (as a lieutenant serving as head of high explosives research in the Navy's Bureau of Ordnance) who hired Einstein.

The deputy director for technical analysis, Blackworld studies division, sat at a table inside the mesh-lined, air-conditioned room 35 feet beneath the deepest roots of the trees that surround CIA headquarters. This is where studies so sensitive they would never see the light of day were kept—the Blackworld. The DD watched the guard—a large man armed with a nasty-looking, large-caliber automatic pistol on his right hip—spin the lock dial on the document safe with practiced skill. The man extracted the HB transcript, brought it to the table, and silently watched as the DD signed the possession card. The guard took the card and then walked smartly into the corridor, sliding the door shut as he left. It locked with a gentle but firm click.

"Push the buzzer on the table, sir, when you're finished," he said through the mesh door. Without waiting for a reply, he walked away down the corridor, his heavy boots echoing in the hall.

There was no need for the guard to stay and watch the DD, as three hidden, closed-circuit color TV cameras kept constant watch over every cubic centimeter of space in the vault. What they saw was simultaneously displayed on monitors in the guard station at the end of the corridor, and temporarily stored on magnetic video tape recorders. No one really thought the DD would surreptitiously try to microfilm the report, but security was routinely tight for everybody. The President of the United States would be treated no differently.

The DD reached inside his coat pocket and extracted a handwritten note. Could the HB tape *really* be as important as the destruction of the M-1 NATO tank? The mystery of how the tank had been blasted out of nowhere, torn open like a gutted fish by superheavy depleted uranium shell fire in the middle of West Germany, had been the DD's assignment. He'd been in on the top-secret study for more than 6 weeks now, and essentially nothing had been accomplished. The pressure was on, and the DD resented anything that distracted his attention. This sudden change in his schedule had better be worth it.

But Henderson had written the note, and Henderson was top-flight. The DD had never known the senior analyst to be wrong. If Henderson thought there was something the DD should see right away, then there *must* be. The DD read the note again.

Allen,

I've transcribed the HB tape myself; not even my own staff has heard it. We're damn lucky it was addressed to the President or it might well have been leaked. Something like this, if it should go public, would bring all the UFO nuts out in full strength. And I can only shudder at how Pentagon hawks could use this to "explain" the M-1 tank ambush. Lucky for us, too, it was

Professor Matthew, one of our free-lance historical scholars, who discovered it misfiled among the Hitler Bunker records over at the National Archives. Matthew is an historian and wouldn't know a photon from a phonograph, even if he did try to listen to the tape (which, for technical reasons, I doubt he could have).

The audio quality is pretty awful—the recording level must have been low to start, and the oxide coating on the tape has been rather badly damaged. Probably due to the casual tossing around it got in its years over at the NA. Still, using some of the nice gadgetry you've kindly let me buy these past few years, I've managed to make out very nearly all of it.

I have lightly edited the transcript into a form suitable for immediate reading by the Director. Where I thought it would be helpful, I've added commentary in the form of "Analyst's Notes." Our friend starts his message immediately with lots of ranting and raving. Later on, he calms down. I'd venture he's an unstable person, at the least, but Jacobs in the psycho-personality section can no doubt provide you with insight on that score.

My formal evaluation and recommendations are at the end of this transcript. Call me if there is anything more I can do.

Gordon H.

The DD replaced the astounding note in his coat pocket and turned his attention to the report on the table. It was thin, only a few pages thick, with a heavy white cover. It was tersely labeled (in bright, red letters) **TRANSCRIPT OF THE HB TAPE**, and carried the routine security legend "Eyes Only" above the Agency's imprint. He vaguely recalled the tape's discovery being discussed at the last directors' meeting—it had been misfiled in a National Archives collection of historical records concerning life in the Hitler Bunker just before the Führer blew his brains out—but it hadn't yet been processed. The DD fished a slightly bent cigarette out of his shirt pocket, lit up, and opened the report. He shook off the childish feeling somebody was behind him, peering over his shoulder—the damned, eerie room with its hidden TV cameras gave him the creeps—and began to read.

D. B. Cooper is a skunk, a rotten, no-account thief! We were going to split the two hundred grand right down the middle. His half because he was taking all the chances. My half because it was my brain that made the job possible. But he left me high and dry! Took it all and ran. I'm telling you this because I want the case reopened, The FBI has given up, you know. Not officially, but they aren't **ever** *going to catch him without help. Once you hear old D. B. is still alive, and how he did it, I hope you'll track him down and throw his butt in jail for fifty years. Here's how it happened.*

Up until last year I was working as an electronics technician for a Boeing subcontractor in Seattle. I didn't work on any of the mechanical design of the 727, but I hung around with the fly-boy engineers who did, so I learned a lot. Enough for the job, anyhow.

[Analyst's note: It was a Northwest Airlines 727 that D. B. Cooper hijacked on Thanksgiving eve, November 24, 1971.]

My specialty is analog circuit design, and if I say so myself, I'm damn good at it. I never graduated from college, but I took all the math and electrical engineering courses I could get. I got all A's in all of 'em, too. It was the university requirements in the non-technical stuff that did me in—I flunked History of Western Civilization **three** *times, and I think that set a school record. Anyway, I got into trouble with my supervisor because I was fooling around on company time with old Wallingsford's equations. Now, don't go running to the library because you won't find 'Wallingsford' in the index of any book. Huntington Wallingsford was a retired scientist who stayed at my university as a visiting lecturer in physics the year before they booted me out on academic suspension. That was two years before I met D. B.*

[Analyst's Note: Professor Wallingsford is still mostly a mystery, but I have been able to unearth one bit of intriguing information about him. He was, in the early 1940's, a civilian English scientist associated with a classified Navy project. The project was cancelled when a destroyer escort, the USS Eldridge, mysteriously vanished at sea and was never recovered.[3]]

Everybody else there thought the old coot was sorta eccentric, but I kind of liked him. I could see the old guy was really sharp, even if the other students didn't. I made it a point to talk to him when I had the time, and he seemed to appreciate the attention. One day he showed me how much. "Mr. [name missing]," he said to me, "I'm going to show you something you won't believe! Come into my office, please." I followed him in and watched as he unlocked a file cabinet and pulled out the dirtiest, oldest-looking notebook I've ever seen.

"Here it is, my boy," he said, "my **life's** *work. I've spent thirty years on my equations, and I've finally solved them! Here, look at them, look here!"*

The old geezer looked like he was about to pop a gasket, so I looked. There was page after page of the neatest handwriting I ever saw. At first, I sorta understood it. It started off with a non-linear traveling wave equation, and then there were Maxwell's equations for the electromagnetic field, and one or two other things that seemed familiar. Then he combined them somehow into really complex-looking stuff and after that I was lost.

---

[3] Official Navy records show that a destroyer escort named *Eldridge* survived the war intact. But was that ship the *real Eldridge*, or was it a secret replacement to cover-up the real ship's disappearance? The Analyst's Note would seem to indicate that the CIA knew the real *Eldridge* had indeed vanished—anyway, that was (as I wrote the story) the conclusion I hoped *Analog* readers would arrive at!

All the while I was looking at his work, old Wallingsford just stared at me. When I looked up, he asked "Well, what do you think? Can you build it?"

"Build it?" I said, "build what?—I don't even know what the hell this stuff **is**!"

He looked surprised at first, and then smiled. "No, no, of course you wouldn't, how could I be so stupid? It has taken me thirty years to figure it out, so how could you understand in a mere few minutes. But you **will**—*soon!"*

"And you know what, he was right! Over the next three, maybe four months, I met with him in his office after classes and we'd go over the math. What old Wallingsford had was the theory for what he called a 'photon duct,' a gadget for changing the trajectories of photons of any frequency, including those of visible light."

[Analyst's Note: Our friend offers no technical information concerning the details of his machine. Suffice it to say his claim is at best **improbably possible**, *and we must emphasize we can offer no explanation ourselves as to how it might be done. However, as Maxwell showed a century ago, light is just a particular form of electromagnetism, and so, assuming a 'photon duct' is even possible, then it is plausible it would be electronic in nature.]*

"The only hang-up was, while Wallingsford was good with pencil and paper, he was the original klutz in the lab. He could just walk into a lab and something would break, burn out, or melt. He needed somebody with the golden touch with electronic circuits, and he'd heard of my reputation in that department. Yeah, why be modest—old Wallingsford needed **me**!"

Once I understood the math, I tried to build the circuits. Man, **how** *I tried. And no dice. I just couldn't get them to work. Wallingsford would come into the lab and hover around, offering suggestions, twiddling knobs, and busting things. Sometimes he'd mutter stuff like "It* **must** *work. It somehow worked on the held ridge. It should work now!" When I asked him what he was talking about, and what the 'held ridge' was, he shut right up. Said he'd tell me 'the whole story' when the time was right, but he never did.*

We worked on it for months. The worst of it was it **should** *have worked. There were some pretty weird circuits, but it* **should** *have worked. We were putting a lot of electrical energy into the power circuits, and it was going somewhere, but the thing didn't get* **hot**. *And no energy was radiated. But nothing* **happened**, *either. We had* **something**, *but we couldn't figure out* **what**!"

Well, that's the way things went the whole year. Then I got suspended and Wallingsford retired for good and dropped dead of a heart attack three months later. All I had was the notebook and a pile of electronics that ate electricity. Still had to eat, though, so I got the job in Seattle. But I didn't give up. When I get my teeth into an electronics problem I don't give up. Now maybe it wasn't really proper to use company time, and the company electronics lab, or the parts bin in the lab, but the problem was like an addiction. I **had** *to work on it.*

So that's the way it went for **another** *no-luck year. Then one d ay, when even I was ready to throw the thing in the trash, I tried something in the equations that neither Wallingsford or I had done before. By then I was willing to try* **anything**. *I went through the equations, and everywhere there was a square root I used the* **negative** *root, not the positive one. When I got done I had a new machine, but it was weird because only a few circuits got changed and even then only just a bit. But that was all it took.*

I turned it on, and I swear—the sucker disappeared! You could reach in with your hands (and watch 'em vanish right up to your wrist), and you could **feel** *the hardware. There was sort of a bubble around the thing. It was an invisibility cloak!*

[Analyst's Note: The term 'cloak' is most appropriate. However the device might function, assuming it can even be built, the net effect would be that of what would be called a 'spherically symmetric surface mirror.' That is, a ray of light impinging on the 'bubble' must not be reflected, but rather caused to skim along over the surface and to depart from it at the exact point the ray would have reached (and at the appropriate angle, of course) had the ray traveled along a straight line. To an external observer, it would be just as though the ray **had** *traveled in a straight line, with no detour. How this astonishing effect might be accomplished, of course, I have absolutely no idea. But there are two other effects that logically follow from the premis that such a device exists. I'll mention one now, and leave the other for later. Both seem to be fatal flaws. The first is rather subtle. All energy radiated* **inside** *the bubble must* **remain** *inside, or else it would be detectable. This means a human body inside such a bubble, with its energetic metabolic processes, must cause the interior temperature to rise. To remain inside such a bubble would soon become intolerable, just as would wearing a heavy winter coat on a warm summer day. A 'cloak' it would be, indeed.]*

So picture it, will ya. There I was, my fingers caressing this invisible pile of electronics, with my hands missing, and all the months of frustration over. I just let out the biggest whoop you ever did hear. And in walks my shift supervisor. He was so ticked off he didn't notice my hands were gone. I quickly stood up with my body between him and the photon duct and flipped the power off. The duct—and my hands—snapped back into view. I was lucky—he didn't see a thing. To make a long story short, all my freebie work had finally caught up to me and I got canned. I was told to clean out my stuff by the end of the day.

I couldn't take the photon duct out past the plant guards, but that was okay. I had the wiring diagrams burned into my brain. I ripped apart the unit on my bench, right down to the transistors, so nobody would accidentally stumble onto what it could do. I lost my security clearance, too, but I didn't care. I guess I should have, considering all the trouble that's happened since. Anyway, even though I was nearly broke I went out and got loaded, just to celebrate. My only regret was that old Wallingsford wasn't around to tie one on with me.

It was at the bar I met D. B.

We got to talking and drinking, or maybe I should say **I** *did, and D. B. mostly listened. By the time the bar closed early next morning D. B. had heard all about the photon duct. I don't* **recall** *telling him, but that afternoon, after I'd forgotten all about it, I got a call from him. He was willing to foot the bill for a new duct, but he wanted something in return. If I was interested, I should meet him back at the bar.*

Well, I **really did** *want to build another duct. I went back to the bar and D. B. laid it all out for me. He'd put up the cash I needed to build a miniaturized duct—about two grand—***if** *I helped him with his plan. Which, of course, as you well know, was the hijacking of an airplane.*

[Analyst's Note: On November 24, 1971, Northwest Airlines flight 305 was boarded in Portland, Oregon by a man calling himself D. B. Cooper. The B-727 took off for its last stop—Seattle, Washington—with a crew of six and thirty-six passengers. Minutes later Cooper gave one of the flight attendants a note demanding a skydiving sport parachute and ten thousand twenty-dollar bills—or he would blow the plane up. To lend credibility to his threat the man displayed a briefcase containing two large red cylinders and a great deal of wire. The plane landed at the Seattle-Tacoma International Airport, the authorities met Cooper's demands, and the plane took off again. Now on board were just one flight attendant, Cooper, and a cockpit crew of four. Three Air Force jets followed in pursuit, but lost the plane in a snowstorm. After being told the plane had insufficient fuel to reach Mexico, Cooper had the plane headed for Portland at an altitude not to exceed 10,000 feet and a speed of 200 mph. He then locked the flight attendant in the cockpit with the crew. At 7:50 P.M. a control panel light signaled the plane's rear boarding ramp was unlatched. At 8:10 P.M. a second light indicated the ramp was fully extended. At that time the plane was over the Cascade Mountains of southern Washington, near Lake Merwin. The outside temperature was seven degrees below zero. The wind chill factor, of course, dropped the temperature considerably lower. Hours later the plane landed at Reno, Nevada. D. B. Cooper and the money were gone. A massive, four-state search failed to discover a trace of either. No chute, no body in an impact crater, no money. Absolutely **nothing** *was* **ever** *found, with the exception of a small amount of rotted cash years later.]*

I'm not making any excuses. I knew it was a crime. But the way D. B. explained it, no one was going to get hurt. He wouldn't have any sort of weapon on him, and no explosives. It was going to be a very 'low-key' job. That's why he only asked for two hundred grand, instead of ten million bucks! We charged only what we felt the market would bear without it getting desperate and pulling guns. Also, $200,000 in

$20 bills only weighs 22 pounds, which he could carry. But I was dumb, too. I should have known D. B. wasn't going to be happy with just that one job. So, once he got off the plane, why, he just kept on going and took **all** *the money* **and** *my photon duct!*

[Analyst's Note: The briefcase the authorities believed to be a bomb might have been the photon duct machine, The two large, red cylinders, made to look like explosives, might have been powerful lead-acid batteries to serve as a portable power source. I emphasize the speculative nature of these 'might-have-beens.']

Get the picture **now***? See why the cops could never find any trace of D. B. after he jumped out? Sure, it's because he* **never jumped***! Once the plane landed, and before the FBI busted in to free the crew, D. B. turned the duct on and made himself and the money invisible. D. B. lowered the tail boarding ramp—that's one of the things I learned about the 727 while working in Seattle, that it could fly with the ramp extended—just to set up a red herring. He was supposed to throw a small bundle of money out on the chance it would be found and everybody would figure he'd been killed in the jump. It hasn't been found yet, but it's out there.*

[Analyst's Note: After our friend recorded this, as I noted two Notes back, a money bundle **was** *found.]*

Before the initial confusion died down, the bum just walked out of the plane, right down the ramp, and past the cops! He got away clean! Take a look at the record since then and you'll find a lot of heists over the years that could only have been pulled by an invisible man. It was D. B, **for sure***, every time.*

[Analyst's Note: I have conducted an extensive computer search of the historical crime records. Our friend is correct in that there have been crimes easy to explain by postulating an invisible criminal. But, of course, there could be other explanations, too, not requiring such an extreme assumption. The description of Cooper casually striding off the plane brings me to the second fatal objection to the possibility of a photon duct device. Just as no internal energy can escape the bubble, then no external light can penetrate it. Indeed, this is the whole idea of the 'photon duct.' But, of course, if no light can enter the bubble, then how could D. B. Cooper see? In fact, anybody inside such a bubble would be **in the dark***, totally blind!]*

So, there I was, fat, dumb, and shafted. I couldn't go to the cops and tell 'em what happened. They'd have tossed me in the slammer along with D. B., if they ever caught him. I couldn't even go to the military brass and beg forgiveness. It **is** *a nutty-sounding idea, I'll admit that, and without a working model to demonstrate right away, they'd have kicked me out on my fanny. Once they saw I'd had my clearance pulled for goofing off on the job, they probably would have turned me over to the cops as a crazy.*

And anyway, I ain't giving the duct away for free.

Old D. B. has really done a number on me. I **want** *that guy! Help me nail the bastard, and maybe we can work a deal on the duct. I'm not much on politics, but if everything else is equal I'll talk with you guys instead of the Russians. Otherwise ....*

[Analyst's Note: At this point there follows a rambling sequence of instructions on how to make initial contact. The idea seems to have been for the President to place coded items in the personal sections of various daily publications on certain specified dates. Of course, he never did that because the tape went astray.]

END OF TRANSCRIPT

The DD leaned back in his chair, and rubbed his eyes hard with his knuckles, Was this the start of a new silly season, he wondered. That time of the year when all sorts of weirdos and kooks surfaced with their deranged theories explaining the secrets of the universe? God, what utter nonsense this HB file is! The DD felt a momentary flash of anger at Henderson for the time lost from the tank study to read this dribble out of some fool's unbalanced mind. But no, Henderson was right. This sort of inflammatory, outrageous crap was dangerous precisely **because** of its potential appeal to the more hysterical elements.

The DD snuffed out his third, stale cigarette since he'd entered the vault. He made a face as he smashed it into the ash tray—the damn thing had made the vault smell like one of brands out of Eastern Europe made from camel dung. He pushed the buzzer for the guard and, while waiting to be released, quickly scanned through Henderson's final summary page. It was consistent with the tone of the Analysts' Notes throughout the transcript: D. B. Cooper was dead, a decomposed corpse lost in a mountainous wilderness, the tape was a fraud, the photon duct was rubbish, and the whole business should be sealed off and forgotten. The DD agreed without reservation, slapped the report shut, and stood up as the guard arrived. The man entered, carrying a small aluminum box with a meter. The meter needle was pulsing.

"Do you have any electronic devices on you, sir?" asked the guard.

"No," replied the DD, startled by the strange question. "Why do you ask?"

"At the same time that you entered the vault, we began to pick up some kind of regular interference on the monitor screens in the surveillance room. A periodic, static-like bursting in the screen image. We thought there might be a malfunctioning piece of electronics on you that could be radiating into the TV pick-ups here in the vault. This box will let me do a quick body scan, and if there is something like that then we won't have to rip into the camera system."

"Be my guest," quipped the DD with a smile, and then he spread his feet wide and held his arms over his head. "Scan away."

The guard slowly moved the box along the DD's legs, upper body, and arms. The needle continued to pulse rhythmically, but no more than it did when the guard held it anyplace else in the vault. "Well, the problem must be in the TV system, itself," said the guard. "Sorry to have held you up, sir." The DD waved the guard's apology aside with another smile and left the vault.

Later that afternoon, the DD sat in his office looking at photo enlargements of the shot-up M-1 tank. Thank God nobody had been in it! He'd hoped for inspiration, but none had come. The entire affair was as deep a mystery as ever, and when the telephone buzzed it was a welcome break from his frustration. His secretary informed him it was Henderson.

"Yes, Gordon, how are you?"

"We're in trouble, Allen. Big trouble." The analyst's voice sounded strange to the DD—strained, nervous. A voice totally out of the ordinary for the normally calm scientist. "Can you come to my office—right now? It's urgent!"

The DD was a man sensitive to the projected emotions of others. Henderson was on the verge of panic. "Yes, of course, I'm on my way."

Henderson intercepted the DD in the corridor and almost pulled him into the inner office. The DD noticed Henderson was sweating, and his hands were trembling. Fascinated by the out-of-character behavior, the DD watched as the analyst practically yelled at his secretary, "No interruptions, no calls. None!" and locked the door. Henderson faced the DD and said "I know you've read the HB file."

The DD nodded and replied "I agree with your assessment completely. At best it was a smart-aleck college prank. At worst it could cause a lot of trouble. It's locked up for good." Henderson was shaking his head vigorously.

"No, I was **wrong**, Allen, wrong! Do you understand what I'm saying? My God, our most secret facilities are **wide open**! Of course, now you've got the answer to the tank puzzle."

The DD stared at Henderson in amazement. The man was making absolutely no sense. "What the hell are you talking about? You, yourself, gave good, solid technical reasons why this so-called photon duct is bunk."

"Allen, I neglected to make the most obvious check, an **easy** check, on the credibility of the transcript. I didn't check the newspapers for the days when our boy wanted to see the President's reply printed—because obviously there'd be none." Henderson sat down behind his desk and looked at the DD. "Well, after lunch I asked the microfilm librarian to pull the personal sections for those days in the two papers that we carry that were on our friend's list. There

were coded replies in them, just slightly different from the code for our side. The bastard must have sent the other side a tape, too. And they **did** answer."

"Oh, come on, Gordon," said the DD, brushing away the analyst's concern, "that's just the kind of nice little detail you'd **expect** out of a clever prankster."

Henderson stood up and began to pace back and forth. "Look," he said, "I got a call from the Blackworld security guards after you left the vault this morning. They told me about the monitor interference. They also told me it stopped right after you left."

The DD felt the first touch of concern. "What are you thinking—that the other side has somehow planted a bug on me?" The DD glanced down at his shoes nervously. Was he a walking transmitter?

"That's what I thought, too," replied Henderson. "So I got hold of the magtape that was running during the time you were in the vault. The guards said the interference was periodic, and I figured I might be able to get some timing data from the tape, That might help in figuring out what kind of a bug we're dealing with."

"And?" The DD was positive now that Henderson was working up to something a lot worse than some simple bug.

"Look at this," said Henderson, leading the DD over to a table. A TV monitor and a tape playback machine were on it, a tape already threaded in the machine. "This is the tape of you in the vault. When I ran it through the first time I could see nothing of interest except for the burst interference. It was at a far too low a frequency for a malfunctioning bug. So I slowed the tape way down, and synched the monitor display circuits to the interference itself. That stuff is caused by electronic switching, all right. The pulse rate is low, but each pulse is very short in duration. That makes it too fast for the eye to see. My God, it's so obvious, and I missed it!"

"Christ, Gordon, **what** are you talking about?"

Henderson's voice went flat with resignation. "I told you the photon duct was baloney because it would be hot and dark inside the bubble, right? Well, what do you do when a coat is too hot? You unbutton it. What do you do when a room is too dark to see in? You open a window. Watch this." He turned the recorder on, and the tape reels began to turn at a very slow rate.

The DD, horrified, watched himself in the vault, reading the HB transcript while smoking what he'd thought was a too-old cigarette. And right behind him, snapping in and out of view, stood a small man. A man wearing an odd-looking, bulky belt around his waist. The man, his hard, Slavic face intense with concentration, was reading the report over the DD's shoulder.

The man, who was smoking a cigarette wrapped in the distinctive red and black paper of one of the more awful brands from the Ukraine, seemed to be smiling, almost on the verge of laughter. The DD understood why.

6

Digital Pictures

In 1901, the German philosopher and scientist Kurd Lasswitz (1848–1910) published a short story called "The Universal Library."[1] Lasswitz was a gifted writer who, a few years earlier, had published a novel dealing with contact between Martians and Earthmen, and it appeared at virtually the same time Wells published *The War of the Worlds*. It wasn't published in English until 1971, however (under the title *Two Planets*), and I didn't mention it in the interstellar invasion chapter because it isn't an *invasion* story at all, but rather one of peaceful contact. Unlike Wells' murderous novel, the German novel had little influence outside of Germany, but "The Universal Library," by contrast, has had a *lot* of influence, as it describes a surprising concept of deep intellectual significance.

"The Universal Library" is told as a conversation between a mathematics professor and a magazine publisher (the professor's wife is also present) who is looking for an essay contribution from the professor to print in the next issue of the magazine. At first, the professor is uninterested in the invitation, saying "I don't have any good ideas right now. Besides, so much superfluous stuff is being written and, unfortunately, printed too—" To that, the professor's wife says "I have sometimes wondered that you can still find something new for the printer. I should think that by now practically everything that can be expressed by letters has been tried."

That's an opening for the professor to introduce the amazing concept of the *Universal Library*. As he says, "The number of possible combinations of a

[1] It appeared as part of a short story collection, and was later translated into English by the German-American science writer Willy Ley (1906–1969), a translation you can find in Clifton Fadiman's 1958 *Fantasia Mathematica* (immediately following the translation in Fadiman's book is a Postscript by Ley that elaborates just a bit on Lasswitz's story).

© The Author(s), under exclusive license to Springer Nature Switzerland AG 2026

P. J. Nahin, *The Art of the "What-If?"*, https://doi.org/10.1007/978-3-032-15618-1_6

given number of letters is limited. Therefore, all possible literature must be printable in a *finite* [my emphasis] number of volumes [the *books* of the Universal Library]." The professor then develops the straightforward mathematics that calculates the size of the Universal Library, starting from a few simple assumptions.

Imagine, he asks, that we assume the availability of 100 different characters for the letters of the alphabet, the numbers, and a variety of punctuation marks.[2] Assume further that a *book* is defined as a volume of 500 pages, with 40 lines per page, with 50 characters per line. Thus, a *book* has (500)(40)(50) = 1,000,000 characters in it, each of which could be any one of the 100 characters that are available. So, the number of different books in the Universal Library is immediately given by

$$(100)(100)(100),,,(100)$$

where there are a million of the (100) factors. That is, the number of books in the Universal Library is

$$100^{1,000,000} = \left(10^2\right)^{1,000,000} = 10^{2,000,000},$$

a gigantic number (1, followed by two million zeros)![3]

Since these calculations have included every possible combination of characters, the professor states a stunning conclusion: "[We] get all the works which ever have been written in the past or can be written in the future." This so shocks his wife that she exclaims "What? You say everything will be in that library? The complete works of Goethe? The Bible? The works of all the classical philosophers?" "Yes," replies the professor, but he then points out that there *is* one complication. "Lots" of the volumes in the Universal Library will be utter nonsense (one of the books will be nothing but question marks, for example), since every possible combination of characters is present, and there is no way in general to tell if any particular volume is "correct" or has "errors."

[2] There is a *lot* of ambiguity here as, for example, what alphabet are we talking about? English, Chinese, Cyrillic, …? Lasswitz's professor chooses 100 characters to give himself a fair amount of "wiggle room" on this point.

[3] That *is* a big number, for sure, but it is *not* the biggest number that mathematicians have encountered. You might initially think of the number mathematicians call the *googol*. That is, 10^{100} (1, followed by 100 zeros), and then of 10^{googol} (1, followed by a googol of zeros) called the *googolplex*. But even the huge googolplex is smaller than the *vastly* larger number $10^{10^{10^{1000}}}$, called the *second Skewes number*, which appears in a theoretical analysis of a problem in the theory of prime numbers. See my book *In Pursuit of Zeta-3*, Princeton 2023, p. 204, for details.

The Argentine writer Jorge Luis Borges (1899–1986) hinted at the professor's point in the very title of his 1941 short story "The Library of Babel,"[4] in which Borges has a librarian of the Universal Library describe the near-impossibility of finding anything specific in the Library (the *architecture* of the building housing this incredible collection of books is nicely described, too).

The concept of the Universal Library is, when first encountered, a shock for most people. It was, indeed, a brilliant insight by Lasswitz, but it was one that was also limited by the technology of his day. After I first read his story in *Fantasia Mathematica* it soon occurred to me that a technical development that came along long after Lasswitz's death enabled a radically expanded version of the Universal Library. I am talking of the development of *digital pictures*, like the ones generated *by the millions every day* by people taking cell-phone pictures (of their pets, of their kids, of themselves, of the dents in the family car for the insurance company, and ... the list is endless). This realization led me to conceive of *The Universal Photo Album* (TUPA), which is even more stunning than is the Universal Library.

After thinking for a while on how to present the concept of TUPA to the world, I had the answer handed to me by the editors of *Omni Magazine*, the first *glossy* paper (*not* pulp!) magazine totally devoted to science and science fiction. A regular feature of the magazine was a one-page humorous essay as the final page of each issue (every essay was logically titled "The Last Word"). The *Omni* editors asked me to submit something for that feature, and I quickly saw that as the perfect way to unveil the concept of TUPA.

This chapter is the only one in the book not to end with a story (I'll make up for that in the next chapter, which ends with *two* stories), and so I'll proceed here just a bit differently by interspersing commentary with the essay itself (which appeared in the April 1982 issue of *Omni*). To open the essay, I reminded readers of how the *idea* of the Universal Library has entered the popular mind, in the form of a parable about monkeys and typewriters.

"The Last Word"

"Just about everyone who learns how to read eventually comes across the story of the monkeys and the typewriters. The idea is elegantly simple. Merely put a bunch of monkeys in front of an equal number of typewriters (and a lot of paper) and let them bang away on the keys. They'll produce mostly gibberish, of course, but the immutable laws of probability predict they will also

[4] Borges' story can be found in his *Ficciones*, Grove Press 1962.

reproduce verbatim *everything* that has ever been written. Like this issue of *Omni*, this essay included."[5]

"And they'll produce everything else that will and even could be written. All the books of the future will be included in the monkey's output: masterpieces of science fiction, fascinating volumes of 'history' describing and analyzing events yet to occur and a lot that won't, and treatises on marvelous scientific breakthroughs."

[**At this point I introduced Lasswitz's Universal Library, which I'll skip here.**]

"This incredible storehouse of knowledge would lead to tantalizing revelations. If a time machine could be built, somewhere in the megatons of paper spewing from the typewriters will be contained the directions on how to put one together. So would be the recipe for a faster-than-light spaceship drive, a cure for cancer, and the secret of longevity."

"But the rub, of course, is that there would be a multitudinous quantity of chaff with the wheat. How could we search through this literary equivalent of the Augean stable to find such gems?"

[**At this point I introduced the idea of an *automatic* generation of black-and-white digital images by writing a computer program to run *systematically* through all possible images that could be formed on an array of m-pixels by n-pixels, with each pixel brightness (the so-called *grey level*) coded in k bits, with 000…0 and 111…1 denoting the extremes of pure-black and pure-white, respectively.[6] I'll describe how to do that systematic generation at the end of this chapter, something I didn't do in the essay.**]

"What would we get from all this? Simply all possible black-and-white photographs: all that have been taken, all now being taken, and all that could ever be taken. We get TUPA."

"What a treasure it would be! In TUPA would be included pictures of every human who has ever lived—Moses, Jesus, Henry VIII, you, me—as well as every human who will ever live. There will be images of every creature in the universe, some of them more incredible than we could imagine. You'll see in

[5]The monkey/typewriter parable can be traced back to a 1913 paper on statistical mechanics by the French scientist Emile Borel. The idea has since then reappeared numerous times. The English mathematical physicist A. S. Eddington, for example, included it in his 1927 book *The Nature of the Physical World*, as did Sir James Jean in his 1930 book *The Mysterious Universe*. In fiction, you can find the parable in the tale "Inflexible Logic" by Russell Maloney (1910–1948) that appeared in the February 3, 1940, issue of *The New Yorker*, and in the October 1943 issue of *Astounding Science Fiction*, with the story "50 Million Monkeys" by Raymond F. Jones (1915–1994).

[6]If a TUPA image is 3000-by-3000 pixels, for example, we have a 9-megapixel digital image, a typical value in ordinary digital cameras.

TUPA all the vacation photos you will ever take ... Best of all, the Universal Library will be there as well, because TUPA will have a picture of every page in every book ever [or will be] published."

[**At this point, to end the essay, I made a pitch for the Defense Department to fund what I called *Project TUPA*, to perform the actual physical generation of TUPA. I urged that because ...**].

"TUPA would include every top-secret [Russian] military document, present and future. Project TUPA could be our next great endeavor, rivaling the Apollo and Manhattan projects. But we must act now because, as the Russians will soon realize, if they haven't already, TUPA contains photos of every top-secret American military document, too."

There is absolutely nothing *logically* wrong with anything in "The Last Word," and no *physical laws* are violated, either. But the *mathematics* of TUPA .... Well, *that* is something else. Here's why. TUPA is an expansion of the Universal Library by an *enormous* factor (we'll calculate the "hugeness" of TUPA in just a moment), an expansion that contains so much more than does the Universal Library with *fewer* assumptions. The question of alphabet selection, for example, never comes up. That's because TUPA contains numerous, complete renditions of the Universal Library for *every* possible choice of alphabet, including those of civilizations on all the planets in the universe, past, present, and future. TUPA also contains images of all possible 2D-paintings: the Mona Lisa, for example, is in TUPA, in all possible ways, including not only the famous version with her enigmatic smile, but also one of her sticking her tongue out, and another of her eating a hot dog with mustard running down her chin.

This sounds like great fun for sure but, alas, while TUPA is physically imaginable, it will forever be mathematically beyond reach. That's because saying TUPA is really, really big is a vast understatement. So, how big *is* TUPA? As before let's take a black-and-white TUPA digital image as a rectangular array of m-by-n pixels, with each pixel having a grey-level coded in k bits. So, each TUPA image has a total of mnk bits. Since each bit is either 0 or 1, the total number of TUPA images is

$$2^{mnk} = \left(10^{0.30129996}\right)^{mnk}.$$

If $m = n = 3000$, and $k = 8$ (giving 256 possible levels of brightness for each pixel), we get a total image size of TUPA of

$$\left(10^{0.30129996}\right)^{72,000,000} = 10^{21,674,160}$$

images. That's a 1, followed by more than 21 million, six hundred seventy-four thousand zeros![7] Not as big as the googolplex, but a *lot* bigger than the googol.[8]

We can systematically generate all the images in TUPA, without generating any image more than exactly once, by simply taking the binary number 000...0 (there are nk 0's) and counting it up to 111...1. That will, however, take a while! To see that, suppose we had started to generate TUPA images at the rate of 100 trillion per nanosecond (that is, at the rate of 10^{23} images per second), beginning at the instant of the Big Bang (15 or so billion years ago). Then the number of images generated up to now would be

$$\left(10^{23}\right)(15)\left(10^{9}\right)(365)(24)(60)(60) = 4.73\times10^{40},$$

a vanishingly tiny fraction of all the images in TUPA. TUPA has *so many* images, in fact, that to find the particular one that correctly describes how to make a time machine (for example), is *so* unlikely that it might as well be in a fireproof safe at the center of the sun, or inside a black hole. So, while TUPA is certainly a fun concept, it really isn't science fiction but, rather, a *fantasy*.

[7] Add a few more bits to the k bits that code pixel brightness, and we could also then code the *color* of each pixel. The size of TUPA would then mushroom once again, but at this point what's a few more million zeros after the 1, right?

[8] This isn't to imply that the googol is a numeric shrimp. It is, for example, easy to show that the googol is larger than the number of raindrops that have fallen in all the storms that have occurred over the entire history of the Earth. (Assume it has rained 24/7 *everywhere* on Earth for its entire existence of 5 billion years, at a rate of 1000 drops per square inch per second—that's a deluge!) Your result should show that the total number of raindrops is but a mere "drop in the bucket" compared to the googol!

7

Time Travel

Time machines are *the* super-science gadget of science fiction, and time travel (particularly to the past) offers-up *the* mysterious adventure of all adventures. The bottom line is that time machines and time travel present an *ocean* of possibilities for writers of imaginative fiction to sail on.[1] There are three (at least) closely related reasons for this: (1) While no one has ever actually built a time machine, there is no known physics that forbids such a construction (like conservation of energy forbids perpetual motion machines); (2) There *are* known arrangements of certain forms of matter/energy that *do* (on paper) function as time machines; and (3) Time travel is full of apparent paradoxes that seem to reduce the idea of time travel to absurdity, but which *do have* "answers" that many physicists and philosophers think convincing.

Let me now offer a few additional words on each of the above three reasons for the fascination of time machines and time travel in science fiction. I'll keep my comments brief (see the references in note 1 for *much* more) but will provide enough to get you set for the two stories that follow. Starting with (1), let me turn immediately to two authorities, one from physics and one from science fiction (Stephen Hawking (1942–2018) and Larry Niven, respectively). The English physicist *rejected* time machines because he hated the possibility of time travelers being able to fiddle (so he thought) with history. He tried to discover, right up to his death, the *physics* that would forbid time travel to the past, but he never did and was forced to fall back on simply

[1] For book-length treatments on time machines and time travel, see my two books (both published by Springer) *Time Machines* (2nd edition) and *Time Machine Tales*. Both books contain extensive discussions of both the physics and the fiction of time machines, but *Time Machines* leans just a bit more toward the physical side, while *Time Machine Tales* leans just a bit more toward the fictional side of time travel.

© The Author(s), under exclusive license to Springer Nature Switzerland AG 2026
P. J. Nahin, *The Art of the "What-If?"*, https://doi.org/10.1007/978-3-032-15618-1_7

coining a jazzy phrase called the *Chronology Protection Conjecture* (that last word is important). As he was fond of saying, the Conjecture "makes the universe safe for historians."

Before Hawking was Niven, with his similar *Niven's Law* that *asserts* (but does not prove) time machines cannot exist because they would allow the past to be changed. Modern thought rejects the arguments of both Hawking and Niven as being based on a confusion in distinguishing *affecting* the past and *changing* the past. Both men modeled their objection to time travel on a version of the famous argument by the physicist Enrico Fermi (1901–1954) concerning the existence of interstellar civilizations: if they exist, where are they? His answer: we don't see aliens from interstellar space because they simply don't exist. Hawking and Niven ask the same question about time travelers, and offer the same answer.

A simple rebuttal to them was given *decades* ago, before either Hawking or Niven were born, in the story "The Time Bender" by Oliver Saari (1918–2000), which appeared in the August 1937 issue of *Astounding Stories*. As Saari (who was a mechanical engineer in real life) explains in his tale, a time traveler cannot travel into the past to a time before his time machine (based on a superdense mass that warps space time, an idea that routinely appears in present-day physics journals) came into existence. So, if a time machine is built tomorrow, then a hundred years from now a time traveler could use it to journey back to tomorrow, but *not* back to today. The answer to Hawking and Niven is simply that we haven't seen any time travelers (yet) because a time machine has not (yet) been constructed.

Turning next to reason (2), a 1974 paper describing a *theoretical* way to build a time machine, by the physicist Frank Tipler, fundamentally changed the status of time travel. Before 1974, time machines were simply fun imaginations, but (*of course*) they were really just absurd super-science gadgets. After 1974, however, time machines became at least semi-respectable.[2] Tipler's opening sentence shocked readers: "In short, general relativity suggests that if we construct a sufficiently large rotating cylinder, we make a time machine." Tipler's paper certainly caught the eye of Larry Niven, who used the title of Tipler's paper for the title of a short story that is based on his Niven's Law: "Rotating Cylinders and the Possibility of Global Causality Violation" which appeared in the August 1977 issue of *Analog Science Fiction*.

---

[2]Tipler's paper (published in advance of his 1976 doctoral dissertation at the University of Maryland) appeared in the *Physical Review D*, one of the premier journals in the physics community. In his dissertation, Tipler offered a more cautionary tone about the possibility of time machines.

In 1988 the Caltech physicist Kip Thorne and his students discovered a new way to build a time machine, out of a space-time structure called a *wormhole*. Soon after, in 1991, the Princeton physicist Richard Gott discovered yet a third, totally different approach to creating a time machine, based on the behavior of mass/energy distributions called *cosmic strings* (see note 1 for more on the time machines of Tipler, Thorne, and Gott). None of these machines has the slightest possibility of being constructed by any human technology likely to be developed in (say, the next thousand years), but the point here is that, when it comes to time machines, known physics seems to be saying "well, *maybe* …".

It is with reason (3) that we find the *real* attraction time machines offer to the imagination. *Paradoxes*. Who doesn't love a good, brain-wrenching paradox? And time travel to the past offers them up by the bucket! The big one, of course, is the famous grandfather paradox, which dates from the early days of science fiction. Imagine a time traveler who journeys back to when his grandfather was a baby and then kills him. The paradox is in the question: *how* did the traveler come into existence so that he could carry out his murderous trip? This particular paradox is just the best-known version of a class of paradoxes that result from changing the past, and such paradoxes can actually be found in pre-twentieth century stories (see note 1).

An entirely different class of paradoxes is those created by a *causal loop*. The classic example of such a loop tells of a young physicist who one day is visited by an older version of himself; the visitor stays just long enough to give the younger version a handwritten (in ink!) lab notebook describing, in great detail, how to build a time machine. Years later, after finally having built the machine, he uses it to travel back to give the lab notebook to his younger-self. The paradox, of course, is the issue of just where did the construction instructions come from? Who wrote those carefully inked notebook pages? It certainly wasn't the physicist: the inked lab notebook was handed to him, he kept it in his possession all through the years it took to build the time machine, and then he handed the notebook to himself. Not once did he put pen and ink to the notebook. So, who wrote the notebook?

While the use of a causal loop has a long history in time travel science fiction (see note 1 for numerous examples), the loop I just described does suffer from a subtle flaw: despite its seemingly straightforward and simple form, it is *not* self-consistent. See the following box for more on that, and for how an early science fiction story (written by a *lawyer*!) cleverly avoided the flaw (*decades* before the pulp science fiction magazines started).

The paradoxes of time travel *are* fun, and I'll end here by quoting two particularly erudite writers on them, one a philosopher and the other a science fiction writer. As the Princeton philosopher David Lewis (1941–2001) wrote, in a path-breaking paper,[3] "Time travel, I maintain, is possible. The paradoxes of time travel are oddities, not impossibilities. They prove only this much, which few would have doubted: that a possible world where time travel took place would be a most strange world …" And, as you read the next two stories, keep in mind what a time traveler says in explaining time travel and its paradoxes to a skeptic (in Robert Heinlein's 1957 novel *The Door Into Summer*), "It's all logical when you look at it mathematically."

The Problem of the Time Traveling Time Machine Notebook

The problem with the apparently self-consistent causal loop of the time traveling lab notebook is this: both the ink, and the paper on which the ink resides are, when handed to the younger-version of the physicist, *older* than they were when the older version received the notebook. This will, in fact, be a flaw in *any* loop in time in which a physical artifact travels. A clever way to get around this flaw is to replace the time traveling physical artifact with time traveling *information*, as was done in the 1904 novel *The Panchronicon* by Harold Steele MacKaye (1866-1928). There we read of how Shakespeare, when temporarily suffering from writer's block while writing a play, overcomes it: he encounters a time traveling fan from 1898 who whispers the missing lines (lines she had memorized in her literary club meetings!) of the play into the playwright's ear! So, he writes the play, she memorizes it centuries later, she then travels back in time where she whispers the lines in his ear … . There is no ageing flaw around this loop, with *information* making the trip, but it does leave open an intriguing question: is Shakespeare a plagiarist? Of himself?

I tried my hand at a time travel story several times, and I'll show you two of them here. The first one, "Old Friends Across Time," appeared in the May 1979 issue of *Analog Science Fiction*, and it illustrates how I *eased* into writing such a tale; it has time travel *implied*, and there are no paradoxes. Rather, with it I was attempting to excite a romantic, mysterious feeling of "wouldn't it be wonderful to time travel," while avoiding confronting any deep philosophical questions.

Now, writing a time travel story without including at least one paradoxical situation might seem to be, in some sense, "deficient." I have to tell you, though, the master storyteller who *founded* the time machine sub-genre of science fiction, H. G. Wells, himself, avoided paradoxes in *The Time Machine*. There *are* two places in that work that *hint* at paradox, but Wells elected, both

[3] "The Paradoxes of Time Travel," *American Philosophical Quarterly*, April 1976, pp. 145–152. Reading Lewis' paper was a revelation to me, and it inspired my books in note 1. An excellent book-length treatment of time travel paradoxes is by the philosopher Ryan Wasserman, *Paradoxes of Time Travel*, Oxford 2018.

times, not to pursue the matter. In the opening passages describing the dinner party at which the Time Traveller (he is never named)[4] attempts to convince some friends of the possibility of a time machine, one of his friends observes "It would be remarkably convenient for the historian. One might travel back and verify the accepted account of the Battle of Hastings, for instance." To that another guest replies "Don't you think you would attract attention? Our ancestors had no great tolerance for anachronisms." For his part, the Time Traveller says nothing. The second hint at paradox occurs near the end of the tale, when a disheveled Time Traveller returns from AD 802,701, and beyond. Startled by his rough appearance, one of his friends exclaims "What *was* this time traveling? A man couldn't cover himself with dust by rolling in a paradox, could he?" Again, the Time Traveller offers no response.

If avoiding paradox was good enough for Wells, then I felt I was following in his footsteps with "Old Friends Across Time." When you finish reading it, I will tell you a bit more about Wells and of his personal view of time travel (that may surprise you), which will explain the Time Traveller's silence.

"Old Friends Across Time"

As I sit here in my study, with the photographic evidence spread before me, I can barely comprehend what my eyes tell me must be so. The evidence is incontestable. And yet—I still struggle to believe. Let me try to explain—possibly in the process I will manage to put my tumbling mind to rest.

For as long as I can recall, old photographs have fascinated me. To page slowly through collections of historical pictures, no matter what the theme, was consummate joy. Even when I was quite a small boy, I used them as my time machine into the past. They took me up and away from the problems every youngster has while growing up, and let me wonder of people and places long since returned to dust. Mathew Brady's Civil War photos had a particularly strong attraction for me, with the horror (and yes, I will admit it, the *fascination*) of war frozen in the images of young men dead before life had really begun. To look at the fallen youth of more than a century before, and to wonder who they were, what they had felt and thought—it all sent shivers through my romantic mind.

I suppose I might have become a professional photographer. But somewhere along in the process of looking at pictures, I became aware of the

[4] In the 1960 film *The Time Machine* (starring Rod Taylor) the Time Traveller *is* named, or at least his first name is given as *George* (almost certainly as a tribute to Wells' middle name).

miracle of the *technology* of picture taking. That led me to chemistry and optics, and finally by some wondrous route, I became an electrical engineer. I never lost my love for old pictures, though, but merely turned my interest in them to the photographic history of electrical physics.

To search for and acquire (for by now I had started my own collection) a photograph of Steinmetz, smoldering cigar clamped in his mouth, giving a lecture on AC circuit analysis using the then still mysterious square root of minus one, made my heart beat faster. To find a faded picture of Einstein at a long-forgotten conference, caught forever in time with his quiet, gentle eyes looking into mine, would send me to the heights of ecstasy.

But it was Maxwell that led me to my incredible discovery.

There is no doubt that James Clerk Maxwell was the greatest theoretical physicist of the nineteenth century. Together with Einstein, he was the best of *any* century. Could it possibly be more than mere chance that the *same* year saw the death of one and the birth of the other? It was Maxwell who gathered all the then known (but fragmented) bits and pieces of knowledge about electricity and magnetism together, and stirred into that mix his own mysterious contribution of the displacement current. There was no experimental evidence then to justify that last step, but the genius of Maxwell knew it *had* to be. And then, from his soaring mathematical insight and physical intuition, he took it all and wrote down the four magnificent equations for the electromagnetic field![5]

No one who has seen and understood those beautiful equations can come away without a quickening of the pulse and a flush of the blood. They are not long; you can write all four vector differential equations on the back of a postcard, but oh, what they tell us! With them, Maxwell showed light was electrical in nature, predicted radio waves *two decades* before Hertz discovered them in the lab, explained energy propagation in space, and radiation pressure, and laid the scientific basis for today's television, radar, lasers, giant electric motors, generators, transmission lines, and—well, why go on? The equations are the work of a level of genius we may not see again for a millennium. We have hardly begun to discover the marvels wrapped inside the electromagnetic field equations. With their aid, and that of quantum mechanics, the very secret of life, itself, may someday be unraveled.

---

[5] I indulged in a bit of historical license here, as Maxwell's field equations numbered 20, not 4. It was the English electrical engineer Oliver Heaviside (1850–1925) who, using the then new mathematical notation of vectors, reduced the equations to 4. Heaviside was, along with gazing at old photographs, one of my youthful obsessions, and I eventually wrote a book-length biography of him (*Oliver Heaviside,* The Johns Hopkins University Press 2001), which has the technical scoop on both what Heaviside did with Maxwell's equations and on what Maxwell's *displacement current* is all about. You'll meet Oliver in the next story.

And so I searched for old photographs of Maxwell. He died at his family home in 1879, before the art of picture taking was barely 40 years old. But I knew in my heart that somewhere there *must* be photographs, yet undiscovered, of such a great man. Anyone who has seen the best examples of prints from wet-glass collodion negatives knows they are, in faithfulness of their rendition of detail, better than what we commonly expect today. Working against me was the fact that the wet-glass process was slow, laborious, and unforgiving of mistakes. The taking of a picture was not a minor decision in Maxwell's time. But still I searched.

I searched for one photo, in particular. When Einstein died, a famous picture was taken of his office, just the way he left it for the last time. On the blackboard behind his desk are the last thoughts he had in his long quest for the Unified Field Theory. The writing on the papers covering the desk is clearly legible, and with modern blowup methods, easily readable. At the time of his death, Maxwell was the Einstein of his times. Surely, I reasoned, a similar photograph of Maxwell's study must have been taken. Even though none has come down through the decades to us, it *must* exist! Gathering dust in an old attic trunk, or buried in a long-forgotten album, it had to be somewhere. I vowed to find it.

I began by writing to all of Maxwell's living relatives,[6] asking that they search through family holdings for any pictures concerning Maxwell that they might possess. For the most part all were cooperative, even though more than just a few thought I was somewhat deranged. Still, it was in vain. I did receive a few old pictures never before seen by other than the family, including a poignant one taken in 1901, showing Maxwell's grave in Parton Churchyard at Glenlair, Scotland. A forlorn, wintry scene, with only what seemed to be three men in the far distance, it brought tears to my eyes. Alas, there were no photos of Maxwell's study. But then late last year, while on a business trip to London, I stopped off for a few hours at the historical archives maintained by the British Institution of Electrical Engineers.[7] On a chance, I looked through their massive files on Maxwell and was rewarded within the hour! What I found will haunt me throughout the remainder of my life.

There it was, stuck through its border with a rusty pin, between two pieces of yellowed paper covered with what appeared to be some simple, rough lecture notes. An ordinary looking photo of a study. Obviously overlooked

---

[6] More historical license here. Maxwell married, but the couple had no children. When I use the word *relative*, I am thinking of in-laws, cousins, and intimate family friends.

[7] This was indeed the name when the story was written, but in 2006 the British IEE became the British Institution of Engineering and Technology (IET). When writing the Heaviside biography (see note 5), I did indeed make such an actual photo-hunt pilgrimage to the old IEE (in 1985).

through the years, or at best unappreciated for what it was, it was the almost illegible, penciled notation on the back that convinced me of my find: just a date, November 9, 1879. Exactly 4 days after Maxwell's death, precisely when some unknown yet inspired person (a family member, a neighbor, a local scientist?) would take such a picture! I am ashamed to admit it, but there was no hope the Institution would let me have the picture. And there was no time to order a copy of it, for I was to return home to America that very night. No, that's not true. The *real* reason for what I did was simply that I *had* to have that original, *old* photo.

So, I took it. It was my scholarly undoing, for that dishonorable act destroyed the picture's tie to verified, legitimate historical records. But *I* know what I found is true.

I could barely control my wild emotions on the flight home. Several times I removed the picture from my briefcase, and looked with fascination at the papers lying on Maxwell's desk, and at the tightly written lines of mathematics on the blackboard in the background. My hands trembled with what can only be called lust. Once home, reunited with my well-equipped photo lab, I would learn every secret hidden in that picture.

There are no words I know that can convey the thrill I felt as I began the processing of that priceless photo. Alone in my lab, with all the modern equipment a well-off amateur can buy (a Caesar Saltzman 8 × 10 enlarger with mercury vapor point light source and a 10 × Plane Achromat Nikon enlarging lens), I carefully cropped and blew-up selected views of the blackboard and desk. Printing the enlargements on ultra-fine grain AGFA Brovira paper, I could scarcely restrain myself from peering at them with a magnifying glass as I waited for them to dry.

Then, at last, I had them spread out across my study desk. I tried to force myself to examine each slowly, carefully, in turn, and not to skip from one to another like a child let loose in a candy store with a dollar. The first three were of the desk papers, including what seemed to be a diary. It must have been lost after Maxwell's death since no trace of it exists in the historical records. I experienced a stunning thrill as I gazed upon the scrawled words, but as they were not easily read at once, I moved on. It was the sixth blow-up, of the upper right corner of the blackboard, that sent reeling back into my chair, an equation that shouldn't, no, *couldn't*, be there. But it was.

To understand my reaction, there is one astounding thing you must realize about Maxwell's equations. When Einstein turned the world of physics on its head in 1905 with his famous paper "On the Electrodynamics of Moving Bodies," all the old ideas about absolute motion and simultaneity of events went out the window. Even Newton's laws of motion had to be modified. But *not* Maxwell's. His equations, just the way he published them in 1873, are the same ones studied today: they need *no* relativistic corrections.

How can that be, you wonder, as they predate Einstein's paper by 32 years? The mystery of this has bedeviled the experts down through the years. Oh, they have an explanation, alright. They say that all of electromagnetics is actually relativistic phenomena to begin with, and the laboratory work of Faraday, Ampere, Henry, and the other great experimentalists were studies of relativistic electron interactions in matter. Thus, it is only "natural" that Maxwell's equations need no correction. So goes the "expert" explanation, but it isn't right! I know Maxwell knew about relativity, and understood it perfectly. He knew all about time paradoxes, and the equivalence of mass and energy.

Because how else can you explain the equation visible in my enlargement: $E = mc^2$?

Why, you must wonder (just as I did), didn't Maxwell publish this remarkable result: relativity theory? At first, I believed it was because of a lack of faith in his results. Who would have believed any of it in those Victorian times, so sure of its absolute view of nature? I thought of how Newton, nearly 200 years before Maxwell, had suffered from a similar hesitancy when he wrote the *Principia*. There, when explaining his theory of gravitation, Newton did *not* employ his new invention of the calculus (which he *had* used to make his discoveries), but instead fell back on laborious arguments based on the accepted mathematics of algebra and geometry. Who would have believed him, otherwise?

But then I realized that couldn't be right. Maxwell was a strong man intellectually, and he wouldn't have held back for fear of disbelief. No, it had to be that he discovered relativity and the mass-energy law just before his death, with no time to make his work known. I was still wrong. It was later, when I returned to the enlargement of Maxwell's lost diary and read those painfully cramped notes, that I learned the truth. What I saw there showed me Maxwell had thought long and hard about his final discoveries, and had purposely withheld them. For clearly visible, after I had slowly deciphered the writing, were the following words:

I have seen monstrous events. My blood has run cold at the sights of two great cities leveled to the ground, their inhabitants cruelly put to death, instantly, or left to die slowly from a strange, lingering disease. Other trips, further on, have shown me the root of all these evils is the mass-energy equation, a result I first believed to be my crowning glory. It will be my crown of thorns unless I ban it from my very being. Another will discover it for himself, but my soul shall be free! I have dismantled my machine, and shall never look upon or think of those horrible scenes again.

This passage was dated just 1 month before Maxwell died a savage death from cancer. The reference to "two cities" can only be that of Hiroshima and Nagasaki. His own death was surely caused by lingering too long among their atomic ruins.[8]

Think what this *means*. Quite simply, Maxwell knew the secret of time travel! But even more incredible is that it must be *easy*, if one only knows how, to build a time machine! Think about it: Maxwell had no gigawatt power stations at his disposal, no high technology machine shops, or nanosecond computers. He was not a gifted experimentalist, and once he had predicted radio waves from the field equations, for example, it took others 20 years to generate them. And yet, *he* built a time traveler. Somehow, with just the puny power sources available to him, and a limited mechanical capability, he wrested the *simple* implementation of a time machine from his dynamical field equations.

Yes, yes, I know what you must be thinking. How can I really conclude such an incredible thing from a single equation on a blackboard, and a few words written by a man dying a painful death? A man, clearly suffering dearly, possibly not in complete possession of his once marvelous mind.

That very evening the last bolt of evidence slid into place. Attempting to escape from the emotional maelstrom into which I had fallen, I turned to my old love of picture gazing. I took down from my library shelf a tattered (yet cherished) volume of the Meserve Collection of Lincoln pictures. My slow paging stopped when I came to the famous photograph by Alexander Gardner of Lincoln's second inauguration. This incredible picture shows John Wiles Booth looking down on Lincoln from behind a buttress high on the steps of the Capitol, while below in the crowd are the 5 men who, 41 days later, conspired with him in the assassination.

The following page demonstrated the extraordinary quality of Gardner's work, as it showed an enlargement of Booth's face in which the circular line between pupil and the white of each eye is sharp and crisp! This impressive picture fascinated me, and I wondered if I could create a similar enlargement. It was then I remembered the old picture of Maxwell's grave, sent to me from Scotland, and the three distant figures in the background. They would present my photo lab skills with a challenge, and the effort would distract my mind.

I finished the enlargement just 20 min ago. Those faces! Two of them I can now finally accept as being there—it must have been a pilgrimage for one, and for the other, it couldn't have been anything but a mocking, ironic

[8] Maxwell *did* die from abdominal cancer, at the same age his mother was when she died of the same type of cancer. Maxwell's cancer was almost surely the result of faulty inherited DNA.

gesture. But I wonder if the youngest one really knew who his two companions were? I don't know the answer to that—yet. But there they are, two men with faces my years of historical photo study have made as familiar to me as my own. One is a youthful Albert Einstein. The other, with the signs of death clearly written across his features, is James Clerk Maxwell. The face of the third man is familiar, too, for the third man is me!

Oh, I am a bit older in the photo than I am now. But it's me, alright. A peculiar, jagged scar across the left cheek, a mark from a childhood accident, is sharply visible, and I can run a finger over my face and match the scar perfectly with its image in the enlargement. I'd say I'm about 45 or so in it, no more than 10 years older than I am now. That doesn't leave me much time to keep my appointment, does it? I don't know, right now, how I am going to do it, but I've got to rediscover Maxwell's secret of time travel. I'm sure I'll succeed—after all, there I am in the picture. Somehow, I'll be going back to pick James and Albert up so we can have our picture taken. Ten years—not much time.

I'm really looking forward to meeting my two new friends from across time.

Just before you started to read "Old Friends Across Time," I promised you some elaboration on H. G. Wells' personal view of time travel. Much to the surprise of many fans of science fiction, he didn't much like the idea! We have his own words in support of that claim, in the form of an essay he wrote for *The Cornhill Magazine* in 1945, just 13 months before his death. The essay takes the form of a criticism of *The Time Machine*, supposedly written by a "Wilford B. Butterrave" (who was actually Wells!). Titled "A Complete Exposé of This Notorious Literary Humbug," the essay describes *The Time Machine* as

> [A] tissue of absurdities in which people are supposed to rush to and fro along the 'Time' Dimension. By a few common tricks of the story-teller's trade, Wells gets rid of his Machine before it can be subjected to a proper examination. He cheats like any common spook raiser. Otherwise it is plain common sense that a man might multiply himself indefinitely, pop a little way into the future and then come back. There would then be two of him. Repeat *da capo* and you have four, and so on, until the whole world would be full of this Time Travelling Individual's vain repetitions of himself. The plain-thinking mind apprehends this in spite of all the Wellsian mumbo-jumbo and is naturally as revolted as I am by the insult to its intelligence.

This passage helps explain the Time Traveller's silence in *The Time Machine* when the issue of paradoxes comes up, a silence caused, I believe, by Wells' inability to "explain" them. We shouldn't look down on Wells on that score because, after all, paradoxes are puzzles that (by definition!) confound essentially *everybody*. A science fiction writer and a philosopher have, in agreement on this point, put it this way (concerning the grandfather paradox, in particular):

> Science fiction writer: "The resolution of [the grandfather paradox] is the key to time".[9]
>
> Philosopher: "[The] apparent possibility of a time traveler being able to do away with both grandfather and himself gives rise to such puzzles that we are forced to question its [time travel] intelligibility."[10]

The fact that Wells could write such a powerful work as *The Time Machine*, even while he had no intellectual commitment to the story's underlying premise, is a testament to his immense skill as a writer of science fiction and fantasy.

The romantic notion of meeting an admired person, long dead, with the aid of time travel, held (and still does) a deep fascination for me. After writing "Old Friends Across Time," it was hard for me to resist returning to the theme again and so, a couple of years later, I wrote "The Invitation." This time the central figure is not Maxwell, but rather Oliver Heaviside (see note 5). This was about the time when my interest in Heaviside turned serious in a scholarly sense, and some years later the result was my biography of him (originally published in 1988 by The IEEE Press). The fictional story came first, however, and it sold to *Omni*. It didn't appear in the magazine, however, as the editors decided to instead include it in *The Fourth Omni Book of Science Fiction*, published by Zebra Books in 1985. When you finish reading "The Invitation" I'll tell you a bit more about Oliver, who *really did* have a connection with time travel!

"The Invitation"

I can still recall my first encounter with Oliver Heaviside. Or rather, I should say with his name, since he had already been dead 35 years at the time. I was in the third year of college, and the professor in my electronics class mentioned him in passing. The name made most of my classmates smile, and I immediately conjured up the image of the jolly, fat movie actor Oliver Hardy who was, of course, a bit on the *heavy side*. It was a juvenile response, I admit,

---

[9] P. Worth, "Typewriter from the Future," *Amazing Stories*, February 1950.

[10] S. Gorovitz, "Leaving the Past Alone," *Philosophical Review*, July 1964, pp. 360–371.

but I began to notice from that time on how more and more frequently the contributions of Heaviside appeared in my classwork as I progressed into the advanced mathematical theories of electrical engineering.

Until recently I was quite unaware, and *could not have had* any idea, how that first encounter with Oliver Heaviside would prove to be fateful. Indeed, for you to best understand the full implications of the tale I'm about to relate, you should know a bit about the strange, enigmatic man of whom history has told us so little.

I'm the only person now alive who even knows what Heaviside looked like in his later years; indeed, I actually have a photograph that shows him to have been a short, vigorous man, his eyes alive with the stupendous brilliance of his massive intellect. And another photograph lies on my desk, too. Seemingly benign, its image has changed my life. It shall take me some time to gather the nerve to tell you of *that* one. So let me first tell you about *him*.

Oliver Heaviside, one of the great mathematical and electrical geniuses of our time, laid the foundations of energy propagation by the electromagnetic waves predicted by James Clerk Maxwell. He also theoretically predicted the existence of the ionosphere (known by some as the "Heaviside layer"). His was second (if not equal) only to that of the almost supernatural intelligence of Maxwell, and even Maxwell overlooked Heaviside's most astounding discovery. But still—Heaviside died a hermit, a near pauper with only a tiny English Civil List pension to hold body and soul together. His official biography is short, impressive, and missing his very greatest accomplishment.

Born 1850, died 1925, Fellow of the Royal Society, first recipient of the Faraday Medal, and nephew on his mother's side of Sir Charles Wheatstone (the father of telegraphy, inventor of the self-excited dynamo, and of the stereoscope). There can be little doubt that the young Heaviside's intellectual gift was nurtured by his famous uncle. Unable to understand his strange methods, the mathematicians of the Royal Society gave Heaviside's work little support, an act of ill treatment for which he never forgave them. His strong personality let him prevail in spite of their ignorance of his superior talent.

Using his novel mathematics, Heaviside discovered the distortionless transmission line that made the miracle of the transoceanic telegraph cables possible. Unfortunately for him, his deep analysis led to conclusions that ran counter to those of Sir William Preece, the Chief Engineer of the British Post Office. Preece, by all accounts a genial but very ambitious man (who suffered from all the drawbacks of that vice), had Heaviside's work censored in the official journals. And so his theory fell on deaf ears in England, and it was left to Professor Pupin at Columbia University in America to successfully show the experimental validity of Heaviside's work. Pupin became a millionaire selling the new technology to AT&T, while Heaviside received nothing. Such a lack of practicality in an engineering genius had puzzled me for many years, but I know now why Heaviside ignored this profitable pursuit, and why his public scientific work *seemingly* ceased, abruptly, in 1912—he was devoting his energies to something far more incredible.

Yes, Heaviside was a great man. But a man of whom no photographs after the age of 43 have survived, or at least that is what history has always thought. A man whose writings are studied world-wide by electrical engineers to this day, but one with no mature image for his disciples to revere. An analogy with Jesus Christ almost suggests itself, but possibly *that* is going a bit too far. No, I'll take that back. Surely you are familiar with the Holy Shroud of Turin, the ancient linen cloth imprinted with the negative photographic image of a crucified man believed by many to be that of our Lord. Well, I possess something of the same for Oliver Heaviside that is even more astounding.

Wait, I can see this is all coming too fast. Let me start from the beginning—then *you* decide if I am a charlatan or rather, as I believe, the luckiest man who has ever lived.

Until quite recently I, as have all others interested in such arcane, exotic matters, believed the only surviving full-view picture of Oliver Heaviside to be in existence is a large portrait painting hanging in the hallowed shrine of the Institution of Electrical Engineers[11] at Savoy Place on the Victoria Embankment in London. It shows a bearded, handsome man in his early 40s. But as to his appearance later in life, who can really say? We have always thought there to be no photographs.

This somewhat mysterious aspect of Mr. Heaviside has always attracted my interest, and as a professor of electrical engineering myself, I've made the collection of Heaviside memorabilia a hobby. I even have copies of his birth and death certificates, sent directly to me from the General Register Office at Somerset House, London. Still, no matter what the burning intensity of my curiosity about Oliver Heaviside, it seemed to me that he had departed this world leaving us with no trace of his stay among the living but brilliant writings. Never married, preceded in death mere weeks by the last of his two older brothers, and buried in a near-forgotten grave next to his parents on the outskirts of the small English town of Paignton, there was no hope of finding even a forgotten family album. I had, therefore, contented myself with nothing more than stumbling across occasional, tantalizing tidbits about his life.

[11] See note 7 again. The name has changed from IEE to IET since I wrote this story, but the address is still the same today.

Indeed, to the despair of my poor, long-suffering wife, I had spent the whole of my last sabbatical leave in England in search of all I could find of Heaviside. But after the passage of more than half a century, even a search at the Mount Stuart Nursing Home in Torquay where he passed his final days revealed no new information. He had faded forever into the dark shadows of time.

So you can imagine my surprise when, some weeks ago, the morning mail brought me an overseas registered letter from the London offices of Groffton, McKee, and St. George, Solicitors at Law. As I finished my second cup of coffee prior to leaving for my 10 o'clock lecture on the design of spread-spectrum radio communication receivers, I read the following astonishing, cryptic message:

As per directions in the Last Will and Testament concerning the Estate of Mr. Oliver Heaviside, late of Torquay in the County of Devonshire, you are bequeathed the following items: (1) a wax-sealed, locked metal box, containing unspecified paper documents; (2) the key to said box. These items may be claimed upon your personal appearance in our offices at precisely eight-thirty on the morning of the 29th of January of this year.

That was just 2 days off! A cable to London verified my unexpected inheritance (I am embarrassed to admit it, but at first I suspected my students of having some fun with my well-known interest in Heaviside), and I hastened to make the necessary travel arrangements. For could there be any doubt at all that I would go?

What a strange happening. A man dead many years before my birth, but one I had long admired and studied deeply. Being blessed with his possessions was truly a miracle from the Lord Himself, one that brought unbounded joy. But also, confusion. *How* had it come to pass? I had always believed Heaviside had died intestate. Clearly there had been a secret will, unknown to all but his lawyers. And then the obvious answer came to me. Most likely he had requested that his papers be presented by the administrators of his estate, after the passage of a suitable passage of time, to whomever had shown through his own technical publications to be a worthy heir to the legacy. Certainly, my own papers had often remarked on my admiration for Heaviside's genius, and that must have allowed me to pass the test.

My wife, understanding my life's passion, offered no objections to the trip. A quick call to my colleagues at the University, the rapid agreement by some graduate students to cover my classes, and I was free. Within hours I had booked passage, and was on my way via the shuttle flight from Boston's Logan Airport to JFK International. I was off to England!

As my plane lifted free from New York, headed for London's Heathrow Airport, I wondered what could be in the box. Possibly some fantastic, previously unpublished technical papers? Papers full of new, incredible results? But that was a long shot, I thought. All of Heaviside's known work had long been gathered into five huge books, the two-volume *Electrical Papers*, and his three-volume masterpiece, *Electromagnetic Theory*. A million and a half words in over 2000 pages! And what conceivable reason could Heaviside have had for suppressing a paper? No, the mystery had to be deeper than that.

Of course, I thought, there *was* the possibility it might be the lost, unpublished fourth volume of *Electromagnetic Theory*! That final manuscript was believed to be stolen and gone forever when, shortly after the radio announcement of his death, Heaviside's house had been burgled. In 1949 a scrap of manuscript in rough draft was uncovered at MIT, but many of the pages were nearly illegible. Enough was readable, however, to show that Heaviside had devoted his final work to combining electricity and gravity—he had anticipated Einstein's search for a unified field theory! My imagination ran wild with that thought. What if the missing manuscript *hadn't* been stolen? What if, instead, Heaviside had secreted it away in the box? My heart pounded with the thrill of the idea. Did I dare even *think* of being that fortunate?

When I arrived at Heathrow it was dreary, overcast, and damp. The total opposite of how I felt inside. Tomorrow morning, I would have Oliver Heaviside's metal box containing, I was sure, stupendous secrets. Acquired not at an action, or obtained through a lucky find in a dusty attic (although I've had both experiences, and they *are* exhilarating), but by the indirect action of Heaviside himself. I slept not a wink that night in my hotel, as my mind raced with anticipation.

Next morning I was up, dressed, and ready to proceed to the offices of Groffton, McKee, and St. George by 7. Far too early since they were only a 20 min walk away. Still, as I need a cane to walk even relatively short distances due to a mild arthritic condition in my right knee, I wanted to take no chances on being late. And I did wish to walk, rather than take a taxi. I have always loved London, but the only way to properly enjoy her is to walk her streets.

I ate an unhurried breakfast in the hotel restaurant to pass the extra time, and briefly played with the thought of trying, once again, to find Heaviside's birthplace. I knew he had been born at home on King Street, now called Plender Street, in Camden Town. But I soon gave the idea up as a search doomed from the start—to find it after a century and the destruction of German bombs would be hopeless. I finished my breakfast, paid my bill, and walked out onto the street. It was 8 o'clock and I was on my way. Mere minutes separated me from the box! My spirits soared. As I made my way among the morning throng of hurrying businessmen and fashionably dressed women, I paid scant attention to even the loveliest of the ladies. None of them could compare with the wonderment I felt was soon to be mine.

I found my destination with little trouble, and paused for a moment while I glanced at my watch. It was 26 min after 8. The letter had said *precisely* 8:30, and possibly the emphasis on the time had hidden meaning. The whole affair was so strange I was fearful of doing anything contrary to my instructions. Should I wait outside for a bit before entering the Victorian law offices in front of me? I decided to be cautious, and waited. I stood among the hurrying crowd pushing around me and took pleasure in watching the large number of children and their parents, all apparently tourists on holiday, happily chattering and clicking away with cameras. I understood their excitement, as my own deep affection for England had grown from such pleasant childhood excursions.

At exactly 8:30 I entered the Law Offices of Groffton, McKee, and St. George. I stood in a brightly lit room, illuminated by the sun streaming through the street windows. Guarding three majestic, ten-foot-high wooden doors was a matronly looking lady, somewhat on the plump side. On each door was fixed a brass nameplate, one for each of the firm's principals. The one for Mr. Groffton, in particular, further indicated that he was a member of that elite group allowed to plead at the bar in the superior courts.

I introduced myself to the lady, completing my presentation with a display of the enigmatic letter that had begun my trip. I was quickly ushered into the presence of Mr. Groffton. A portly, elderly gentleman, he rose slowly from behind his desk and greeted me warmly.

"My dear Professor, you cannot *possibly* realize how I have looked forward to this day! Please, please, come in and be seated. By Heaven, I can't recall being so excited in years!"

Such a welcome from a stranger, particularly a distinguished English barrister, was most extraordinary. My astonished reaction must have been plainly visible as the gentleman hastened to explain.

"My deepest apologies, sir. I can see my words must sound like those of a mad Englishman who has been out in the sun too long. But you see, ever since I was a small boy I've waited for this day. My father was a man of the law before me, and he drew up the papers that have finally brought you here. His only wish, in his final years, was to live long enough to be present this day himself. But the private instructions given to him personally by Mr. Heaviside were emphatic: you were to be summoned precisely one month after your forty-first birthday. My father had to wait but, alas, the reaper who eventually catches us all did not wait."

The sheer strength of Mr. Groffton's emotional words was enormous. My feelings were of total confusion, but he had conveyed something of the magnitude of the events enfolding me. This moment in time, the forces that had propelled me across the Atlantic to these law offices, had all found their origin *decades in the past.*

"I am sorry," I replied, "but I don't understand. How could Oliver Heaviside have asked for *me*? I wasn't born until 1940, fifteen years after he died. Surely what you mean is that he asked for someone with *my qualifications*, which I assume are my publicly expressed interest in his work and my professional training in electrical physics."

"No, no," exclaimed Mr. Groffton, "you are mistaken. Mr. Heaviside asked *specifically* for *you*! He was here for several hours and, naturally, before his departure from my father's office, discreet, rapid inquiry, with the aid of the miraculous trans-Atlantic telegraph cable, was made through legal contacts in America about you. There was nothing in the instructions to indicate *when* your forty-first birthday would occur, and my father wished to see how much longer that would be. You can imagine my father's surprise when informed there was no trace of an American professor with your name, or even, for that matter, of an electrical engineering faculty, at your institution!"

I sat down in a nearby stuffed leather chair. It absorbed me like a drop of water on a dry sponge, and I gratefully accepted its comfort.

"I don't see how this is possible!"

Mr. Groffton did not reply at first, but rather returned to his desk. He sat motionless for some moments and, much to my discomfort, seemingly stared at me. He then seemed to realize the awkwardness of the situation and finally replied. "I must tell you, Professor, that in the beginning my father entertained serious doubts as to the strength of Mr. Heaviside's grasp on reality. The mystery, you see, was that he was leaving his estate to a specific individual who apparently did not exist. Mr. Heaviside became quite angry when my father brought the issue forward, and refused to discuss it. However, since he had paid all expenses in advance, the proper arrangements were made and all subsequent actions have taken place automatically."

I said nothing in reply. What *could* I say?

"And then," Mr. Groffton went on, "something occurred that brought a cry of delight and astonishment to my father's lips. He had, of course, during all the many years since, never put the matter of Mr. Heaviside's strange will from his mind for long. His speculations about the 'mad' Mr. Heaviside often made rousing good talk around our family supper table, and they were better than any of the traditional childhood tales, I assure you, sir! Each year we monitored the faculty roster at your university, Professor. Thirty years ago an electrical engineering faculty was at last created. Each year thereafter your name was absent. Until ten years ago."

"Yes," I fairly cried out, "that's correct! That's when I joined the department as an assistant professor."

Mr. Groffton nodded his head, and then stared at me again. "So it seems Mr. Heaviside was not mad after all. But Professor, surely you must appreciate the most obvious, if incredible, implications in all of this."

I was beginning to form some ideas by then, of course, but I lacked the nerve to voice them aloud. I delayed by asking, in turn, "And what, Mr. Groffton, do *you* think it means?"

For just the shortest moment I had the feeling Mr. Groffton would speak his thoughts. But then his nerve broke as had mine. He leaned back in his chair, spread his hands wide, and chuckled. "But of course *that* would be sheer fantasy! No doubt once you open the box there will be a perfectly logical explanation for the entire affair."

For my part, I was not so sure.

The lawyer stood up and was once again in total control. "If you'll come with me, Professor, we'll take care of the business that has brought you across the ocean."

"Yes," I replied, relieved to have the conversation move away from its previous path. "I am most anxious to see my inheritance."

I followed him to the far end of the room, where against the back wall of highly polished, inlaid wood tiles there was a magnificent, heavy oak table. It must have weighed a ton. But it was what it bore upon its surface that captured my attention. The metal box!

Possibly a foot square, and four to five inches deep, its lid was securely fastened by a flush mounted lock on the front apron, and the whole thing was covered by a thick layer of red sealing wax. My hands trembled as I looked at it, and a chill of excitement touched the back of my neck. Mr. Groffton was now fishing about in a pocket and then found what he was after.

"Ah, here we go, Professor, the key." He dropped it on the table next to the box. It appeared much like the latch key to the door of my wife's prized, curved glass antique hutch. Such an ordinary looking key. But I knew it would fit a most decidedly *extraordinary* lock. The lock to Oliver Heaviside's box. *My* box!

Mr. Groffton motioned me to one of the several wooden chairs that matched the table. "You may examine the box at your leisure, Professor, if you wish. I can not tell you how many times I've wondered at its contents but, of course, I understand your desire for privacy."

I *did* wish to savor my excitement alone, but a veneer of automatic, civilized behavior made me deny my own feelings. Mr. Groffton was a wise, sensitive man, though, and to my secret pleasure he brushed away my silly words.

"No, no, Professor, I have a feeling that what you'll find in that box may prove to be quite extraordinary. After you've seen it, *then* you decide what, if anything, you wish to reveal. And I must tell you that Mr. Heaviside's will contains the requirement that you, *alone*, must open the box. I have to be out of the office for the next hour or so on business, in any case." And then he was gone.

The ponderous boom of the giant office door closing behind Mr. Groffton had barely died away before I was seated before the box. I lifted it, and it seemed fairly light. *Well,* I recall chuckling nervously to myself, *it certainly isn't a hoard of gold bars.* I could hear instead what sounded like paper inside (just as the letter from the lawyers had stated), sliding in the box as I gently shook it. I assumed the red wax, often used in high vacuum experiments to seal tiny leaks, had been used to prevent the humid air of London from destroying the documents inside with the rot of mildew. The wax easily pulled away in large sections, much like peeling the shell from a hard-boiled egg. Soon I had it all off. I placed the box squarely in front of me, inserted the key, and with a gentle click that belied its years of dormancy, the lock relinquished its decades-old grip on the lid.

I opened the box.

Inside lay three documents, each covered with a thin film of the dust from the original air in the box. There were two envelopes, and a thick manuscript nearly two inches deep. The manuscript had a blank cover, and one of the envelopes was unmarked. The other had the words *OPEN THIS FIRST* written across it in ink. As I looked at the handwriting I felt a shock of incredible proportion. I have studied Heaviside's surviving notebooks in the archives of the Electrical Institution at Savoy Place, written with ink of his own making, and I recognized the graceful, flowing style. I must admit that but for the energy I felt from the thrill of the moment I might well have fainted dead away from the excitement.

I tore open the marked envelope with trembling fingers and inside found the following note:

June 17, 1918

Dear Professor:

I can well imagine the feelings you must now have as you read these words. I will not waste effort here with a lot of unnecessary explanation. The other envelope should contain proof enough of what I shall claim (if this box, willed to you, isn't proof enough). If you do decide to take me up on my offer, we will have time enough to discuss it all. I admit to a bit of theatrics with my method for communicating with you, but if I had merely shown up on your doorstep, would you have believed me?

I have discovered the secret of time travel.

But I have only just barely discovered it. There is much I don't know. I can move only in discrete jumps, like a frog hopping from one water lily to another. Also, I am fearful of the possible paradoxes of what I've uncovered, and have been most careful not to create one: Nature might prove rather unpleasant in resolving the issue! Other problems I face are far too numerous to recount here. I know from your writings that I have only a few years left to live. I want to make them count! What I need is a colleague to work with me. I need someone to talk with, to work with, to join me. My limited experiments have allowed me to visit your time (I cannot move forward less, and only very far into the future), all I have learned shows me that you , sir, are that colleague. The scientists of my time are either no better equipped with electrical and mathematical knowledge than I, or are unable to even begin to grasp my work. Your writings have convinced me it must be you to join me.

The manuscript describes my machine. Build it and travel back to work with me. In the text you will find all the details for the instrument settings required to return to my time, tomorrow evening in this very room, only hours from now as I write these words. I will know, then, if you have decided to join me. It will take you some weeks to build the machine, but for me it will seem but less than two days.

This morning, I will take this box to a London lawyer. There it shall remain until you receive it. It is up to you, but I feel in my heart you will come. Now, open the other envelope.

I shall wait for you.

Oliver

I sat immobile, awash in near disbelief! I opened the second envelope. It contained two photographs. Both were very old, yellowed, and stiff with age. One was of Oliver Heaviside, in black and white, with a note written on the back informing me that this was the most recent picture of my correspondent, taken only 2 months earlier. I tried to calm my whirling mind by smiling at the thought that the photo, for me, was actually considerably more than half a century old.

The other photograph had been produced by an instant camera, in full color. The colors had long since faded, but it didn't matter. I could look at my own clothes and see how the image had first appeared. *It was of me, only half an hour ago, standing outside on the sidewalk among the children and their parents, leaning on my cane!* At last, I understood the peculiar emphasis on the precise time of my arrival. It was the appointment time for my portrait, taken by Oliver Heaviside, who must have been mere feet away from me in the crowd. He had arranged and controlled it all, subject to the limitation of traveling in time jumps.

I stared at my picture. Was it decades old, or only half an hour? Was the man who had taken it, at this "very moment," waiting for me, or was he lying moldering in his grave? Or was it both? I shook my head in utter despair at getting it all straight. I took the thick manuscript from the box and scanned quickly through it. It was full of mathematics and electrical schematics, but I could see much that was familiar. There were a few strange twists here and there, but, I reasoned, there would *have* to be. I held in hands the plans for a time machine!

I was, by then, too emotionally numb for any further delay. My movements became automatic. I returned the documents to the box and locked it. I placed the key deep in my coat's inside breast pocket. I picked up the box and left Mr. Groffton's office—once outside I quickly signed for my inheritance with the outer office secretary. I ignored her questions and pleas for me to wait, and said only that I wished for her to thank Mr. Groffton, upon his return, for the great courtesy he had shown me.

I returned to my hotel room where, at last, I lay down on the bed and let my mind spin. Could all this *really* be? The will, the pictures, the manuscript? Could anything deny them? No, nothing.

My journey was done, and my first thoughts were to immediately return home. But then, I realized, there was actually no real reason to hurry. If I accepted my incredible invitation, I would appear in Heaviside's "tomorrow evening" no matter when I started my trip back. I could take my time, to try to sort out the jumble of recent events. *That* was what I should do, I thought, and I made up my mind.

A strange urge came over me to delay my return for a day or so. I decided to once again pay a visit to the final resting place of Heaviside, to pay homage to his grave. Possibly then, I hoped, when once again in the presence of his spirit, I would find my way through the maze of reluctant doubts that still tore at me. I regretted that there would be no time for me to listen to the speakers in Hyde Park, or to visit Regent's Park Zoo, or to feed the ducks and geese at St. James, but a more urgent mission pulled at me. I quickly made arrangements to leave London early the next morning.

The trip by rail to Paignton was uneventful, and I soon found my way to the cemetery where Heaviside lay buried. Little had changed since my last visit some years ago. The walk from the gates, to the grave, is slightly uphill, and I was somewhat out of breath when I finally located it. A thick, swirling ground fog had at first confused me. My right knee throbbed lightly from the exertion, and I leaned heavily on my cane. Fearful of letting the box out of my sight, for even a moment, I had carried it with me.

I stood in front of the gravestone, panting, searching for *something*—some sort of sign to guide me, to help me see what all of this could mean. Gradually I became aware that I was not alone. On the other side of the marker, possibly 10 feet beyond and barely visible in the fog, was an old man, apparently in his late 70s. He was shabbily dressed and his white hair was unkempt and dirty. A tattered sweater was all that kept the damp chill from his body. He was staring at me. When he saw I was returning his gaze, he walked haltingly around the grave and came up to me. He was carrying a shovel, and so I presumed he was a grave digger.

"I say, Gov, I didn't mean no rudeness. It's just that I saw you comin' up the hill here, and when you stopped at ol' crazy's grave, I got curious."

The casual manner of the insult angered me. "What do you mean, ol' crazy's grave? Don't you know who Oliver Heaviside was?"

The man grinned and his rotten teeth made the smile grotesque. "Yeah, sure. I heard he was some kind of genius. But to me and the boys he was just an old crazy. When I was a lad, me and my pals, why we used to give him hell, we did. Sneak around at night, we would, yellin; and shoutin,' givin' him a scare. It was all in fun, and we didn't do him no harm, though." The old man snickered to himself at the remembered, long-ago "good times."

I didn't reply, but merely stared angrily at him. I was familiar enough with stories of how neighborhood boys had treated Heaviside. He was strange to them, and so they had taunted him. I had read accounts of how they had even broken the windows of his house on more than one occasion. That had infuriated me, the thought of such barbaric behavior. No harm, indeed! And here I was face-to-face with one of those delinquents, now grown old and decrepit. The only thing that prevented me from giving him a good tongue lashing was his age.

The old fellow's demeanor was suddenly different, then, and he was peering closely at my cane. At first, I thought he must be offended by my obvious disapproval of him, but I was disappointed when that proved not to be the case. The elderly ruffian looked my cane up and down and stared hard at me. "Where'd you get the walkin' stick, Gov? It's a mighty peculiar lookin' one."

Puzzled at his strange question, I took my weight off it and, still holding the box tightly under my left arm, held the cane up. It *is* a strange cane, I must admit, being made of thick, polished, heavy oak, carved with a curious design with a meaning I've never learned, and topped with a solid brass handle. It's the handle that attracts the most attention, as it's shaped in the head of an extremely unpleasant looking gargoyle. It was a humorous gift from my wife some years ago. It is very ugly, but also very strong and able to easily support my weight when my knee troubles me.

"My cane?" I asked dumbly. "It was a gift. Why do you ask?"

The old man looked into my face closely, as if he had found a long-forgotten picture. "Because I saw that cane once before," he replied. "had to be *that* cane, couldn't be *two* like it. It was the last time me and my pals gave ol' crazy a hard time. We were hiddin' down behind a hedge, yellin; and whopping,' when this stranger we'd never seen before came out of the house after us. He was a wavin' a weird lookin' cane—*that* cane—at us, and sayin' he would bash our heads in if we came back. Seemed like *he* was crazy, too, and meant it, and we never did go back."

He kept staring at me, and then he started to back off. His face was white. "Been a long time, now, and it's hard to remember, but that stranger, he looked a lot like—no, it ain't possible!" And then he turned and ran away down the hill.

I smiled as I watched him disappear into the fog, as I'd see him soon enough again. *He* had been the sign I was looking for!

A sudden, cold gust caused me to pull my coat about me as I stood alone again by the grave. The wind began to blow harder, scattering dirty paper scraps through the rows of tombstones. The light grew faint as the sky filled with swiftly growing clouds, and a misty rain began to fall through the fog. I clutched the metal box tightly to my chest, holding fast to the incredible link between the past and now. Or was it a link between now and my future? The astonishing adventures of the last 2 day had left me dazed. But at last my decision was made. I had much to do and I hurried off the damp, dark earth of the graveyard, carrying with me the ghost of Oliver Heaviside.

Had any man ever received such a legacy—an invitation to experience all of history, and the future, too? An invitation from a man I'd never met? Until now had thought I *could* never meet, but had long ago grown to consider a kindred soul? No, indeed, no, there can be no doubt. I *shall* accept his invitation. Oliver and I *will* stride the corridors of time together!

In 2018 the *Philosophical Transactions of the Royal Society A* published a special issue to commemorate the 125th anniversary of the appearance of Heaviside's *Electromagnetic Theory*. The editors of that issue approached possible contributors around the world with an invitation to write a paper for that issue, with the only condition being that the topic of a paper be on some aspect of Heaviside's work in electromagnetics. I was pleased to accept, and submitted a paper with a title that I knew would be provocative.[12] What that title is referring to is Heaviside's fascination with the electromagnetic radiation spectrum of a charged particle moving faster-than-light (FTL). If the motion takes place in a medium other than a vacuum (like water) then such a thing can actually occur.[13] Heaviside thought FTL motion a possibility *even in a vacuum*, without realizing the effects on causality such motion would have (he wrote before Einstein's relativity was published). I tell you all this so you know that I realize "The Invitation" has some historical flaws. But, after all, it *is* science *fiction*!

Finally, to end on an anticipatory note, I'll say more in Chap. 10 on the "changing the past" concerns associated with time travel. In that chapter the idea of alternate or parallel universes is the central topic, a seemingly outrageous idea that has found some support in the world of quantum physics. But it *first* appeared in the science fiction pulps!

[12] P. J. Nahin, "Oliver Heaviside: an accidental time traveler," which appeared in the December 2018 issue of the *Transactions*.

[13] The radiation spectrum for an FTL charged particle is, today, called *Cherenkov radiation*, after the Russian physicist Pavel Cherenkov (1904–1990), who studied it experimentally in the 1930s (work that won him a Nobel Prize in physics). Marie Curie (1867–1934) had, years earlier, observed the pale-blue glow emitted by FTL electrons in radium solutions, but had not understood the glow's significance. A so-called swimming pool nuclear reactor, which is submerged in water for both cooling and radiation shielding purposes, emits a similar blue glow (take a look at the cover of the October 1951 issue of *Scientific American Magazine* for a famous Oak Ridge National Laboratory photograph of that blue glow). Before Curie, Cherenkov, and submerged reactors, however, Oliver had been there first.

8

Future War

It is, today, almost impossible to comprehend the horrors of personal military combat as it was conducted thousands of years ago. Those were the times when vast armies of men would stand literally nose-to-nose and swing rocks, clubs, swords, and battle-axes at each other. All left standing, in the end, were ankle-deep in pools of blood. It was all pretty awful.

The passage of time did bring *some* changes and so, by the time of the American Civil War, Union and Confederate soldiers were shooting at each other with large-caliber rifles and cannons over distances of several hundred yards. The blood pools and carnage, however, remained.

With the passage of a few more decades, came "progress." With the development of aircraft, combat distances increased to *miles* in the First World War, and to *hundreds of miles* in the Second World War. By the end of that war, the German terror weapons of the V1 pulse-jet flying bomb (an early form of the cruise missile), and the V2 rocket, had strongly hinted that combat distances would soon increase even more: with today's ICBMs, even *thousands of miles* of separation offer no safety.

Other equally dramatic changes in war technology were occurring, as well. Beginning with the First World War, the development of radio marked the start of the integration of electronics into combat. It didn't take long for the combining of electronics and weapons to bring into existence what are called *weapon systems* that, for anything more advanced than a personal combat knife, *required* the use of electronics. That is, lethal weapons and electronics *fused*, and it became ever more difficult to see the dividing line between the weapon and the electronics.

© The Author(s), under exclusive license to Springer Nature Switzerland AG 2026

P. J. Nahin, *The Art of the "What-If?"*, https://doi.org/10.1007/978-3-032-15618-1_8

In the 1970s, during an "interesting" (I use this word in the Chinese proverbial sense) year of employment at The Institute for Defense Analyses in Washington, DC, and later as an academic consultant to the US Air Force, I had a first-hand exposure to fusion. The problem I studied then was how widely distributed arrays of various sensors (infrared, acoustic, radar, optical, pressure, and magnetic), on a variety of platforms (ocean floors, submarines, surface ships, aircraft, and satellites) could be *combined*. Each of all those sensors generated their own decision on what they had detected in some common surveillance volume, with each decision coming with a generally different level of confidence. The question was how to best achieve a *fusion* of all those different sensor outputs to arrive at a single master decision of "what was out there."[1] During that work, I started to wonder about the next (obvious) step in fusion: the intimate combination of electronics with the very body of a human combat soldier. The result was the story "The Man in the Gray Weapons Suit," written in 1979 and which eventually appeared in two book anthologies.[2]

The warrior in the story is a naval aviator operating off an aircraft carrier. Alas, when I wrote the story I had never been on a carrier. My lone naval shipboard experience had been in the late 1960s, as a technical observer of the target tracking performance (under combat conditions) of the 3D phased-array radar on the USS *Long Beach*, a nuclear-powered guided missile cruiser. The ship executed such extreme maneuvers (as a target aircraft flew around the ship at various ranges) that it rolled as much as 30° each way from upright. Yep, quite an experience, but one of no help in writing the story, and so I had to just *make up* what I *imagined* it would be like on a carrier.

In the late 1980s, nearly 10 years after I had written the story, I *did* get to spend almost a week at sea on the supercarrier USS *Forrestal*, as a guest on what the Navy calls a "Tiger Cruise" for family and friends of the crew (my son was an enlisted sailor on the *Forrestal*). Watching one night as F-14 fighter pilots performed spectacular carrier qualification take-offs and landings, I realized that, if anything, the reality of carrier flight deck operations was even more "science fictiony" than I had imagined. To watch an F-14 launch from a

---

[1] You can see how I and a fellow University of New Hampshire colleague (John Pokoski) mathematically modeled this question in our paper "NCTR Plus Sensor Fusion Equals IFFN: or, can two plus two equal five?" *IEEE Transactions on Aerospace and Electronic Systems*, May 1980, pp. 320–337. (IFFN is *identification friend, foe and neutral*, and NCTR is *non-cooperative target recognition*.) If you decide to skip looking this paper up (be prepared for some admittedly dense equations), I fully understand. I'd rather read a good story, too!

[2] *The Future at War: Thor's Hammer*, Baen Books 1988 (Reginald Bretnor, editor), and earlier in *Day of the Tyrant: There Will Be War*, Tor Books 1985 (Jerry Pournelle, editor).

carrier, at night, at full military power, was like watching the *Millennium Falcon* go faster-than-light in a scene out of *Star Wars*.

"The Man in the Gray Weapons Suit"

The warrior gently ran his hands over the cool, smooth flanks of his love. She responded not in any physical way, but still he knew that deep within, under the flawless skin, she felt his presence. With a single flowing, graceful motion, he mounted her. She opened wide and he slipped inside. The fit was narrow, his bulk squeezed tightly against her sides. But to the warrior, it was cozy, snug, warm. The thermal-proof plexiglass cockpit housing slid back down over his helmeted head with barely a hissing of its electric motor.

He inserted the digital communications umbilical cable from his biosensor body box into the female connector on the floor and became as one with Red Striker Five. Snapping the throat mike down against his neck, he called the Combat Information Center on the nuclear carrier. "CIC, this is Red Striker Five. All weapons stores on board, fuel topped off, and set for launch. Give me your ready-status readouts."

"Roger, Red Striker Five, we have you on first launch. For your ready-status: wind over the bow at forty-two knots, sea state at level three, cloud cover starts at five thousand, solid to eight thousand five hundred, and clear from there up. You are initially on weapons tight, with a required visual verify of a bogey before authorized to perform a weapon release. No voice communications in the clear allowed—perform all data transmissions to the fleet on GPS LINK 99. Set your security level switch to antijam position three. Ship radars show a clear screen out to one-five-five miles. Prepare to launch."

The warrior felt the passing of a momentary irritation. *Weapons tight, with a visual verify! Damn!* His air superiority platform could detect, lock-on, and track 23 simultaneous targets out to 300 line-of-sight miles. His frequency-hopping, pulse-Doppler look-down radar could pick enemy infiltrators out of the massive clutter background echo of the ocean, even if they attempted to penetrate fleet defense by coming in 100 feet off the deck at Mach two. His "fire-and-forget" Eagle-six missiles, with home-on-jam radiation seeker-heads, could attack and kill with probability 0.999 at 170 miles, at hypersonic speed. And he had to identify a hostile visually before weapon release! Even with his electro-optical visual aid, that meant a maximum attack range of 10 miles. Maybe 15 if he used the narrow-field zoom lens with auto-video tracking. Damn! It was *his ass* on the line, not those of the political toadies who soiled their pants at the thought of a mistaken ID. Having to close that near to a potentially deadly hostile dropped his survival probability by at least 10%.

But maybe the hunting will be good today. The use of the Global Positioning System satellites, and the GPS antijam, encrypted digital data link meant that something was up. His pulse rate elevated, and the surface of his skin wetted slightly with perspiration. Small biosensors in his body box picked those reactions up and routinely filed them away in the on-board flight computer. The gray weapons platform called Red Striker Five would need to know everything if it was to help the warrior survive.

The flight-deck tractor was already hooked-up to his aircraft, and the warrior waited as the 21 ton Red Striker Five was attached to the steam catapult. Eight hundred feet ahead, just faintly visible in the dim, early glow of daybreak, was the edge of the flight deck. From there it was 93,000 feet straight up to his operational limit and possible death, and 90 feet straight down to certain death. No in-between. Just up or down. He briefly thought of what it would be like, sealed in his cockpit as he made the short, yet long fall into the water, and then imagined the essentially infinite inertia of the 87,000 ton carrier as its stately mass crushed him under at 30 knots. No real need to worry about that, as the fall alone would kill him all by itself, without any help.

His eyes and mind turned to the red-glowing cockpit displays. Soon, as he climbed out of the cloud cover and burst into sunlight, he would turn the night-vision lights off. As the final seconds before launch slipped away, he made the last pass through his checklist. Red Striker Five was ready, a deadly air and near-space machine of precise, electronically guided death. Her companion tensed as CIC warned him of launch, and he braced for the high gee acceleration that would fling them up to takeoff speed. Theoretically, it would take only a bit more than half the flight-deck length. But the warrior had witnessed launches that hadn't worked. His mouth felt dry and his heart pounded as he brought the throttle levers up to 60% of full military power. More than that would dangerously stress the catapult mechanism. Red Striker Five roared her pleasure as she gulped the JP5-nitro mix.

The blast of launch pressed him back into the body contour seat, the details of his peripheral sight faded in a tunnel-vision redout as blood was forced away from his optic nerves, and the edge of the flight deck rushed toward him at incredible speed. And then, as Red Striker Five lifted free of the launcher coupling, the warrior's left hand shoved the throttles forward, his right hand pulled back on the stick, and he accelerated up and away. Pulling a 30-degree attack angle, he rode the twin jet engine exhausts at an initial climb rate of 20,000 feet/min. He could feel the whining turbine shafts under his buttocks, one cheek over each roaring monster. He used the feedback through his backside to even-up the shaft rpms to balance out the engines. Balanced engines gave better handling, less fuel consumption, and minimum vibrational abuse to his spine.

The rear-view closed-circuit screen showed the carrier rapidly falling behind and below him. As he hurtled toward the clouds, the massive ship was soon no larger than a postage stamp, and then, as he flicked on the afterburners, it disappeared in seconds. He was riding a tornado now, each engine thundering out 42,000 lb of thrust. At 2 lb of thrust per pound of platform weight, Red Striker Five was like no aircraft that had ever flown before. Her limits were set by the endurance of the warrior, not by technology.

He punched through the top of the cloud cover with a vertical velocity vector of 30,000 feet/min with the airspeed needle past the Mach 1 line. His only indication of breaking the sound barrier in a climb was the funny behavior of the altimeter; first it lurched upward 2500 feet, and then wiggled its way back to normal. The engine vibrations on his butt, the pounding roar in his ears, and the instant response of his craft to the stick, throttle, and foot controls gave him the same pleasures a passionate woman would have. Except no woman could ever be so perfect.

When Red Striker Five passed 30,000 feet, the warrior returned to normal engine power, as the afterburners wasted too much fuel for just cruising. He nosed the craft over gently until he was in level flight at 33,000 feet, with an indicated airspeed of 1200 knots. He flipped off the night-vision lights, set the surveillance radar to its long-range search mode, and turned his Mark XII/IFF identification-friend-foe beacon transponder on. The beacon could be interrogated by coded transmissions from properly equipped observers, and Red Striker Five would automatically broadcast her mission number, altitude, bearing, and airspeed. Her radar echo would also be enhanced on the observer's radar screen. Failure of an interrogated aircraft to respond to such requests could bring an infrared missile up the tailpipe!

"CIC, this is Red Striker Five. I am at start of mission run. Going now to GPS LINK 99."

"Roger, Red Striker Five, we have a solid track and now going with you to LINK 99." There was a pause, and then "Good hunting, Red Striker Five. Take care."

The warrior heard the strange metallic bang on the voice circuit as the digital link took over—he was now in direct communication with CIC's computer. It would talk to him by audio tone cues through his headset, cues that he could react to faster than an oral command. He was now getting the situation normal cue of a continuous up-down frequency sweep from 1 to 5 kHz, each 3 s. The sound was comforting to the warrior, like the humming of a mother to a baby. He snuggled happily in his seat.

Red Striker Five flashed through the sky, serving as the eyes of the mighty naval fleet miles below and behind. Part of the always-flying combat air patrol, CAP war missions were top priority tasks. A billion-dollar carrier, and its hundred-million-dollar[3] companions of missile destroyers and cruisers, couldn't afford to let the enemy get close enough to launch a tactical nuclear-tipped cruise missile. CAP missions extended the fighting reach of the fleet from the 15-mile range of Second World War naval gunfire to the many hundreds of miles that were Red Striker Five's combat radius.

Ten minutes after launch the warrior heard the audio tone cue in his ears change to a frequency sweep period of 1 s. The first stage alert cue. Three bogeys were beginning to edge onto his cockpit radar screen, and the pre-engagement weapons program in the fire-control computer switched the radar from surveillance to track-while-scan. All of Red Striker Five's radar updates were now being automatically uploaded to a GPS satellite, 22,000 over his head. There, integrated with sensor data from surface, subsurface, airborne, and space platforms, CIC had an accurate, real-time picture of everything that moved in a water-air-space volume of over 2,000,000 cubic miles.

The bogeys were now flashing red on the multicolored radar screen, the computer's visual cue to the warrior to pay attention. His left hand punched the Mark XII interrogate button. Seconds later, the flashing red switched to the steady yellow of a friendly response. Good. But the enemy had been known to spoof the IFF by making recordings of friendly replies, and then retransmitting them when their penetration aircraft detected the interrogation request. The warrior went to the Mark XII's HOARSE GOOSE mode and hit the interrogate button again. Now a new, special coded signal, changed daily, was included that would *inhibit* a friendly from replying. *A reply would be a fake.* But these were friends as their images on the screen changed to green, and the audio tone cue reverted to 3 s. The radar returned to routine surveillance, the tracking computer dumped the stored trajectories of the identified bogeys, and the fire-control computer relaxed its electronic finger on the trigger. Red Striker Five streaked on.

Four hundred miles out, with the radar's constantly searching fan beam picking up nothing, the audio tone cue went into a continuous ping-ping-ping. LINK 99 was calling. The warrior pushed a red button next to cockpit video

[3] The story was written nearly a half century ago, and these numbers should certainly be revisited. I decided to leave them as originally published, for your amusement.

screen that normally displayed a rear view of the aircraft. The screen cleared of the flashing sky and an encrypted digital message streamed down from GPS. LINK 99 was a spread-spectrum, time-division multiple-access channel, with hundreds of users tied into it, on a time-sharing basis. With a spectral width of many gigahertz and a signal level 40 db below the background noise level, there was no way the enemy could jam it, even if he could find it in the infinite electromagnetic spectrum. Unless he was willing to devote the entire electrical output of a 2000 MW nuclear power plant to the task, a highly unlikely event. The decoded message scrolled in milliseconds onto the screen:

> *MULTIPLE HOSTILES AT HEADING 079. SUB-SURFACE ACOUSTIC SENSORS INDICATE VELOCITY VECTOR OF 1450 KNOTS, AT 500 FEET ON FLEET INTERCEPT BEARING. WEAPONS FREE DOCTRINE NOW IN EFFECT. ENGAGE AND DESTROY.*

This was more like it! This is what he had been waiting for! He punched the red button again to acknowledge the message and the audio tone cue changed to a rhythmic 6-kHz pulse, with a period of one-half second. It sounded like the speeded-up drum beat that rowers on Roman warships must have heard, thousands of years ago, as combat was about to begin. Red Striker Five was ready to go to war! the GPS would watch for surface-launched missiles directed at Red Striker Five, and warn the warrior by modulating the pulse rate, intensity, and tone frequency in each ear.

He put the radar in its high-power, burn-through-jam, track-while-scan mode, 5° around both sides of 079, from sea level to 1000 feet. Eleven flashing red spots appeared on the radar screen, in a vee pointed back toward the fleet. They were 620 miles from the warrior's home ship, and 215 miles from Red Striker Five, to his left. The warrior pushed the HOARSE GOOSE button. There was little doubt: the attack pattern was characteristic of the enemy, and no friendlies would roar into the fleet in such a clearly provocative, hostile way. But he was going to kill them, and he had time to be sure. Red crosses appeared, superimposed over the flashing hostiles! They had replied to HOARSE GOOSE, thinking they were being normally interrogated.

Too bad for them, thought the warrior. *Let's go get'm baby!*

The warrior snapped on the power switch for his heads-up display. Normally a clear plastic shield extending from the top of the instrument panel to just below the canopy, it now became a remote projection output of the fire-control computer. Once combat started, there was no time to keep moving eyes and head between instrument display and cockpit windshield. The HUD showed the warrior the radar screen image, Red Striker Five's gun-cannon status, and projections of the proper launch envelope for the Eagle-Six missiles against all tracked targets. The warrior looked at the computed intercept course and saw he was already within attack range. He elected to go in closer.

That was his only mistake.

Hitting the afterburner switch, he lit the tail of Red Striker Five and went into a 60-degree banking turn-dive. As he passed through 20,000 feet he was hitting 1500 knots and accelerating. The titanium/boron fiber skin of Red Striker Five had a dim, yet visible cherry glow. The projected intercept course, computed by the dedicated radar computer operating at a memory-cycle time of 43 ns, kept pace easily. The warrior flipped the red plastic cover off the missile arming/firing switches with his left thumb and threw the leftmost of three exposed toggles. He couldn't hear the cryogenic pumps, but he knew the infrared sensors on the missile seeker-heads were now rapidly cooling down to the 10 °K where they operated optimally in the terminal attack phase. The warrior asked for a launch countdown from fire control to achieve missile intercept at approximately 70 miles range, and large numerals projected on the HUD. As the numbers flickered by, he flipped the middle target-attack toggle to designate all displayed radar tracks as hostiles. When the flashing red zero appeared on the HUD, he threw the rightmost toggle on the missile switches, and the Eagle-Six weapons came off their wing hardpoints, two at a time, from each side.

Fump! The first double-pair streaked off, with each of the four missiles locked onto its own target. Guided by control signals from Red Striker Five's radar computer, they would fly their own way in on the last 2000 m of intercept using their infrared seekers.

Fump! The second double-pair raced out and away, their exhaust trails leaving a crazy, swiftly dispersed pattern. Each missile quickly accelerated to 4000 knots, its body glowing red-hot with air friction. The warrior loved night-attack missions, with the blazing missile skins looking like jewels. But even now, in the daylight, he could follow them for a few split-seconds. Then they were gone.

Fump! The last three missiles launched, two from the right wing, the eleventh from the left.

The warrior watched, fascinated, as thin, spidery purple lines, marking the missile paths, weaved their way on his HUD toward the flashing red hostile symbols. At first the enemy vee stayed on course, but then their electronic warning systems picked up the in-bound missiles. The vee started to break-up, the pattern spreading apart. The warrior knew that some were diving, others climbing, but all were being flown by men as good as dead. An Eagle-Six missile could pull 37 gees in a chase-down maneuver, greater than even Red Striker could take without disintegrating. And the enemy platforms were inferior to Red Striker Five. But the doomed men tried. And the warrior had no pity. They were the enemy!

One after the other, the purple filaments reached out and touched a desperately twisting, twisting, whipping, spinning red dot. And then they both faded from the display. The searing explosion, the vaporizing metal, the carbonized flesh—all were reduced to a quiet decay of glowing, colored light reflected in the cold eyes of the warrior.

The tracking radar computer performed an automatic kill assessment of each strike, looking for the highly characteristic fragmentation pattern of a successful intercept. As a backup, for attempted kills at ranges under 100 miles, a spectrum analyzer also examined the radiation from the explosion fireball, looking for a suddenly enhanced carbon line. The last blaze of glory of an enemy warrior before the mist that was once a man's body dispersed forever. A low-level kill assessment would bring a secondary missile attack, but none was needed. All 11 hostile markers had vanished. The warrior flicked off the afterburners and let his machine coast down to 1000 knots. No need to waste fuel.

And then the warrior felt Red Striker Five shudder, and his surprise was unbounded as he saw fireballs bigger than his fist stream by his cockpit, above and to the right of his head. He'd been jumped from the rear, and was taking high cyclic 37 mm cannon fire! With his attention diverted to the earlier attack, a 12th enemy aircraft had somehow evaded detection. *The bastards must have learned how to defeat LINK 99! Maybe those were decoys I just took out!* As he realized his peril, the right wing took two hits: one at the tip and one on the trailing edge near the wing root. Red-hot, searing metal fragments tore through Red Striker Five's body, and one, the size of a man's thumb, ripped into the warrior's right leg, just below the knee. Muscle tissue, bone, arterial fragments, all mixed with shreds of flight-suit fabric, splattered the cockpit, and blood gushed from the wound. Instrument glass splinters ripped into his body. Blinding pain tore at the warrior, and he would have screamed but for the paralyzing shock.

The warrior knew, just before he passed out, that his survival was out of his control. He retained enough strength to slap the emergency combat palm switch at the side of his seat, and then he rapidly slid into unconsciousness. It was up to Red Striker Five to get them both home.

The palm switch activated the autonomous-combat program in the flight computer. Immediately Red Striker Five examined all biosensor outputs on the warrior's body, determined the presence and location of blood loss, and pumped compressed air to the proper imbedded circular tube in the right suit leg to create a tourniquet. The blood flow slowed to a seeping. Simultaneously, Red Striker Five lit her afterburners, blew away all external weapons pods, and dove for the deck. The enemy warplane followed her down, too close for a missile attack, but well within gun range. It was the enemy's mistake.

Red Striker Five leveled out at 200 feet, moving at 1500 knots, weaving, jinking, humping, in a manner determined by a random number generator in the computer software. Desperately trying to keep those 37 mm fireballs away from her warrior! The enemy pilot was good—but Red Striker Five was better. Hurt by the loss of streamlining from the ragged metal edges where she'd been hit, Red Striker Five was melting at 1500 knots. The enemy was 1000 m behind and closing at 1600 knots. Red Striker Five dropped to 20 feet above the ocean, letting her surface-following radar keep her at altitude. The enemy stayed on her tail. The enemy pilot was *very good*, rough-riding through the near-surface thermals on an attack run.

The two screaming metal bullets raced over the water, with cannon bursts rocking Red Striker Five violently side-to-side. The acoustic shock wave each was dragging along was incredible, and a boiling wake of dead fish bobbed to the surface long after the hunter and the hunted had passed. And then Red Striker Five fought back.

When the enemy was only 700 m behind, Red Striker Five popped her air brakes and lost 300 knots almost immediately. Simultaneously, she pulled into a climb and did a full inside loop, coming down behind and on the tail of the snookered enemy aircraft. The defeated foe had only a few milliseconds to realize his fatal error, and then Red Striker Five ripped him apart with two dozen strikes from her dual 20 mm cannons. The flaming enemy debris flared out along a ten-mile track, but by then Red Striker Five, bearing her dying warrior home, was gone.

Racing for altitude, she climbed to 5000 feet and started squawking on all clear broadcast channels:

EMERGENCY–EMERGENCY–EMERGENCY
RED STRIKER FIVE CAP
WOUNDED PILOT ABOARD
REQUEST PRIORITY LANDING
REQUIRE FLIGHT DECK MEDICAL

Over and over she transmitted her urgent message as she bore in toward the fleet with her burden. The warrior flickered in and out of consciousness, but knew neither where he was or what his fate would be. He put his trust in Red

Striker Five and passed out. She didn't fail him. The flight deck was cleared and, with guidance signals from CIC's computer, Red Striker Five made a perfect landing. The Navy medical personnel gently lifted the warrior's torn body from the shattered cockpit and placed him carefully on the deck. After emergency aid, as they prepared to take him below for permanent surgery, he temporarily regained his senses once more.

"Take it easy, son," said the medic, "you're hurt pretty bad, but you'll be okay. I saw it all on the radar screen—that's some aircraft you got there. She fought her way out and back like nothin' I ever seen!"

The warrior smiled weakly through a pale white face lined with pain and shock. He looked up at Red Striker Five and saw not a technological marvel of electronics, armament, metallurgy, and computer programs. He saw both a warm and loving creature, and a being that had killed to save him. Killed with savagery and intelligence. His body filled with emotion, a feeling of passion that only later he would just barely begin to understand. He looked at the battle-ravaged Red Striker Five and, just before he slipped into darkness again, he knew. He knew she'd be there when he came back. She'd wait for him, and he loved her.

And he knew she loved him, too.

9

Topology and the Fourth Dimension

One of the more popular story ideas in science fiction is that of *spaces* with dimension greater than 3. With just those few words I am willing to bet you immediately thought of the *fourth* dimension. In science fiction the word *dimension* has more than one interpretation, of course (in Wells' *The Time Machine*, for example, the fourth dimension is *time*), but in the stories of interest in this chapter the fourth dimension is a *spatial* dimension. The idea of higher-dimensional spaces has broad appeal, and even children are amused by it. In the *Superman* comics, for example, kids soon encounter Mr. Mxyzptlk (whose home world is five-dimensional), and Donald Duck's immensely rich uncle Scrooge McDuck (whose wealth is so vast it is stored in a vault with a six(?)-dimensional volume measured in *cubic acres*).

The enormous popularity of Edwin Abbot's 1884 *Flatland* shows that going in the opposite direction (from 3 dimensions down to 2) has its attractions, too, but it is with the fourth dimension and higher that the *real* adventures begin. The pulp science fiction magazines were, from their start, filled with such tales (see my books cited in note 1 of Chap. 7 for numerous examples). Many of those early stories routinely started with a character getting the reader up-to-speed with some introductory talk about what mathematicians call a *Möbius strip*.[1] All science fiction fans know that the strip has the curious property of being a surface with just *one* side, as is easy to demonstrate. Suppose the strip is made of white paper. Put a red crayon on any point of the strip and start coloring *without lifting the crayon from the paper*. Keep coloring for as

[1] Named after the German mathematician August Möbius (1790–1868), who described it in 1858. The strip is easy to make. Just take a long, narrow strip of paper and stretch it out flat on a table top. Then take one end of the strip and give it a half-turn twist. Then, finally, tape or staple the twisted end to the other end. An illustration of the strip is at the very end of this chapter.

© The Author(s), under exclusive license to Springer Nature Switzerland AG 2026

P. J. Nahin, *The Art of the "What-If?"*, https://doi.org/10.1007/978-3-032-15618-1_9

long as you see white paper. Eventually you'll see that there is red *everywhere* and no white paper is visible *anywhere*. Without the half-twist you can't do that without lifting the crayon and turning the strip over to get access to the other side. That's because the *un*-twisted strip is two-sided, while the half-twisted strip is one-sided.[2]

The ease of making a real Möbius strip was fascinating even for non-mathematical readers, and one of the very best stories based on the strip appeared not in a science fiction pulp magazine, but rather in the once very popular general audience magazine *The Saturday Evening Post*.[3] This story involved the painting of a twisted drive belt on a piece of machinery and is quite clever. In contrast, the pioneering days of pulp science fiction featured tales not nearly as sophisticated as Upson's. A "good" example of those early stories is "The Captured Cross-Section" that appeared in the February 1929 issue of *Amazing Stories*. Written by Miles J. Breuer (1889–1945)—who was, in real life, a practicing doctor—it is about a mathematician who loses his girlfriend in the fourth dimension when he builds a machine that can rotate four-dimensional space. That sort of story has its charm, yes (especially for teenagers), but lacks the "gravitas" for which "dimensional" science fiction later became famous.

Good examples of what I am thinking of can be found in some of the stories by Homer C. Nearing (1915–2003), in which two college professors embark on what would ordinarily be thought absurd adventures. Appearing over a dozen times in *The Magazine of Fantasy and Science Fiction* in the early 1950s, and featuring math prof Cleanth Penn Ransom and philosophy prof Archibald MacTate (think Holmes and Watson, respectively), each story starts by Ransom laying the mathematical groundwork for the rest of the tale. (MacTate keeps things moving along by regularly asking Ransom for clarifying explanations.)[4] Nearing was not a mathematician (he was a professor of English at the Pennsylvania Military College), but his stories were told in a highly sophisticated manner which showed he was impressively well-read in mathematics.

[2]This amazing property inspired the well-known popularizer of mathematics, Martin Gardner (1914–2010), to take the next step to a zero-sided object, with his famous story "No-Sided Professor," which you can find in *Fantasia Mathematica*.

[3]"A. Botts and the Möbius Strip," by William Hazlett Upson (1891–1975), in the July 1945 issue of the *Post*. Upson was not a mathematician, but rather a former salesman for the real Caterpillar Tractor Company. His story character, Alexander Botts, is a salesman for the fictional Earthworm Tractors (you write best about what you know)!

[4]I so admired this writing technique that I copied it in my technical book *The Mathematical Radio*, Princeton 2025. There I have two math professors, Tweedle and Twombly, discussing the laws of electricity. That was a lot more fun to do than just putting down a lot of equations.

In "The Hermeneutical Doughnut," for example, you'll find one of the best discussions that I can recall reading of the famous map coloring theorem, and of its quite different results for maps on a sphere and for maps on a torus. Another tale, "The Hyperspherical Basketball," is a hilarious description of playing basketball with a ball that can enter the fourth dimension. Finding Nearing's tales, today, is probably not easy, but I was able to purchase (from a used book dealer on the Web) a 1954 collection: *The Sinister Researches of C. P. Ransom*. If you can find this, it will provide you with a fun romp through academia *and* mathematics.

The best-known tales in "dimensional" science fiction are, I believe, "A Subway Named Möbius" by the American astronomer A. J. Deutsch (1918–1969) which appeared in the December 1950 issue of *Astounding Science Fiction*, and "—And He Built a Crooked House" by Robert Heinlein which appeared in the February 1941 issue of the same magazine. (Both tales were bought by the legendary editor John W. Campbell who, you'll recall from Chap. 1, also published "The Cold Equations.")

Deutsch's story introduced its readers to the topic of *topology*;[5] the mathematical study of objects stripped of all distinguishing characteristics other than their *shape*.[6] In the story we are told that the Boston subway has been relentlessly expanding, with the continual addition of ever-more stations and their connecting tunnels. Then one day, with the addition of yet one more station, a train "goes missing." A Harvard mathematician soon comes forward with an explanation: The train has "hit a node, which is not [a physical] obstruction." Rather, "It's a singularity. A pole of high order."[7] The mathematician goes on to explain that the subway complex already had a "high order of connectivity" and then, with the addition of the new station, "the connectivity has become infinite."

There then follows much "mathematical talk" in which the mathematician "explains" how, while the missing train is still in the subway, it is inaccessible. As he says, "The train has no real "whereness." The whole [subway system] is without real "whereness." "It's double-valued, or worse."" The mathematician admits that topology is not his real research field, and that the subway people

[5] The word *topology* is due to the German mathematician J. B. Listing (1808–1882), who had independently discovered the Möbius strip a few months *before* Möbius.

[6] The classic joke about topologists is that, while undeniably smart, they nevertheless can't tell the difference between a teacup and a doughnut. After all, each is simply a compact blob with a single hole in it and so, therefore, they are the same thing!

[7] Readers who have studied the mathematical theory of integration in the complex plane, and of the need when doing such integrals to avoid integrating through a singularity of the integrand (which will be of some order), will recognize where Deutsch found his motivation for using these words. As far as the storytelling goes, of course, all that actually matters is that the "tech-talk" sounds "sophisticated."

should call-in an expert for definitive help. Luckily, such an expert is in Boston, a professor at "Tech" (clearly a reference to MIT). At first, it looks like the expert won't be needed as the missing train reappears, but then another one disappears—and the Tech professor is on it! We are then told that there may be no way to bring the subway connectivity back down to a finite value and, on that somewhat dark note, the tale ends.

As a way to combat that darkness, just a bit, I suggest reading Deutsch's story while listening to the Kingston Trio's famous 1959 song "M.T.A." (of Charlie, "the man, who never returned" after getting on a Metropolitan Transit Authority train in the Boston subway system). That song, written in 1949 as part of a Boston mayoral campaign, might well have been the inspiration for the following year's appearance of Deutsch's tale (in 1949 Deutsch was a lecturer at Harvard, and so he certainly heard the song more than once)!

Heinlein's tale is that of a Los Angeles architect who constructs a house in the form of a four-dimensional cube (sometimes called a *tesseract*) that has been "collapsed" into our normal three-dimensional space. Then, the not-uncommon event (in Los Angeles) of an earthquake causes the house to "collapse" back into four-dimensional space. As the architect explains what has happened to his somewhat perplexed client, "this house, while perfectly stable in three dimensions, was not stable in four dimensions. I had built a house in the shape of an unfolded tesseract: something happened to it, some jar or side thrust, and it collapsed into its normal shape—it folded up."

The story continues with a guided tour through the collapsed house, with much explanation of the oddities of four-dimensional geometry. Heinlein's description of the collapsed house is, I think, perfectly correct, but it helped that in its original publication in the magazine some explanatory artwork was included with the story, artwork which generally is not reproduced whenever the story is reprinted. The only technical flaw (in my mind) is when the architect says the house was stable in three dimensions but unstable in four—I think the reverse makes more sense (but who am I to criticize Heinlein!).

I have long been fascinated by stories of dimension/topology and it was inevitable that, at some point, I would begin to wonder, now-and-then, if I could write such a tale. I only tried to do it once, but that was one of my rare experiences where a first attempt at a story-type *succeeded*. That tale was "Twisters," which appeared in the May 1988 issue of *Analog Science Fiction*. In it a higher-dimensional version of the Möbius strip (called the *Klein bottle*)[8] appears, in a rather fiendish form.

[8] Named after the German mathematician Felix Klein (1849–1925), who described it in 1882.

"Twisters"

For years I had walked by the empty corner lot on Fifth and King Street, over on the north side of town, every day on my way to work, and that's just what it had always been—a vacant corner lot. Local kids might build a snowman on it in the winter, but the rest of the year it was pretty much ignored. As the years went by the weeds changed, and maybe a rusted, abandoned car would give way to some discarded furniture, and once there was a pretty big fire there when somebody torched a pile of half-rotted auto tires; but really, there was never anything of consequence in that lot.

Then, one day last week, right out of the blue, there was a doughnut shop in the middle of it. That puzzled me at first because it hadn't been there the day before, and I hadn't seen any construction work going on. It seemed strange to me, this doughnut shop appearing overnight. But then, I told myself, with modern building techniques and materials they could probably put up a hotel just as fast.

The shop was attractive, in a strange sort of way. Most doughnut shops are just square, white buildings with large plate glass windows. If it's a new shop, the windows are clear, but gradually a thin film, probably grease, forms on the glass. If you see dead flies stuck on the glass, that is a sure sign you definitely don't want to buy a doughnut there! This place wasn't like that at all. Instead, it looked like an old-fashioned antique shop, something like the curiosity shops you see as illustrations in Charles Dickens' stories. I could tell it was a doughnut shop because that's what the sign said, and the smells coming out of the shop were irresistibly fantastic. So, as an accomplished connoisseur of doughnuts, in I went.

I must have been the first to discover the shop because there was nobody else inside. At first, I thought the shop might not yet be open for business, but it smelled *so good* I didn't leave and instead decided to stay and look around a bit. Along the walls was a glass display counter filled with doughnuts. There were some ordinary looking doughnuts of the usual round shape with a hole in the middle. No surprise there. But then I saw big stacks of chocolate ones, with such a curious twist in them that I couldn't focus my eyes on a whole doughnut at once. I shrugged that off, assuming it was the result of some special optical effect caused by the way the glass of the counter reflected light. As I thought about those strange-looking doughnuts, I heard a discreet cough, and then a voice from behind me.

"Ah, sir, may I help you?"

I turned and there was a little man with long, pointed fingers, a flat nose squished over half his face, and bright, red eyes. Yeah, that's right, *red* eyes and *real* bright, which seemed almost to glow. I remember thinking the fellow must have gone without sleep for days, rushing to get his shop built and stocked, to have eyes that bloodshot. He smiled at me—with teeth that came to points so sharp it looked as though a rough file were jammed sideways in his mouth; but he seemed pleasant enough. I remember thinking, though, that he really ought to see a good dentist.

"Yep, I think you can," I said in reply, sticking my hand out in greeting. It's not a big town where I live, and everybody knows pretty near everybody else. "Hadn't seen your place here before," I went on, "and thought I'd stop in and introduce myself. I live just a couple of blocks up the street and walk to my office, every day, right by this lot—ah, I mean your shop. I am the local M.D. here, by the way. And from the way your shop smells, I suspect I'm going to be a steady customer."

The little fellow took my hand in both of his, the way those European dignitaries greet each other on the evening news, and uncovered those meat grinder teeth of his in a big grin. "Well, thank you, doctor, I appreciate so much your friendliness. Every customer is a special customer, that's my motto." As he talked, he rubbed my hand—not offensively, but I had the feeling in the back of my mind—didn't really think about it until after I'd left, in fact—that he was massaging the skin on the back of my hand, and feeling the fleshy part where the thumb connects. Nothing *real* strange, mind you, but definitely not a standard handshake, either.

"We are running a one-week special, doctor, to build our visibility quickly in the local community." The clerk was literally beaming now as he spoke, with a big, toothy smile. "With the purchase of a dozen of our regular doughnuts, another dozen of any other doughnut style is absolutely free."

Well now, even though I do yell at some of my patients about losing weight, I guess I'm a sucker for all the wrong things, too. There is nothing I like better than taking a fresh doughnut and slopping it into a cup of hot coffee! Some people think doughnut dunking is just one notch above eating soup with your fingers, but for me it's one of the supreme pleasures of life, and I guess I gave him a big hint on that score when I replied "Now that's pretty darn nice of you. I think I'll take you up on that. I'll have a dozen of those chocolate twisters." I pointed to the pile of doughnuts I had looked at—*tried* to look at—before. "Tell me," I went on, "how come your special doughnuts have such an odd shape? Don't get me wrong, they smell wonderful, but they sure don't look at all like your regular ones."

The man chuckled and rubbed his long, bony hands together. "Now doctor, that's our trade secret, you see. We do it different from any other doughnut shop, certainly, to stand out, but the shape is also the key that gives us a *much tastier* product, by far. Oh, yes, indeed it surely does!" He then walked quickly around to the back of the counter, scooped up my two dozen doughnuts in waxed paper and popped them into two big paper bags. The regular ones went into a white bag, and the chocolate twisters were sealed in a bright blue one.

"Here you are, doctor. I know you'll have a new eating experience with these, and I look forward to—*meeting*—you again."

I thanked the fellow and left. Even if he was a bit woods queer, he sure knew how to make a fragrant doughnut. When I got to my office, Millie—she's my nurse—greeted me with a big hello and a mug of steaming black coffee. Millie knows my passion for dunking and when she smelled the aroma coming out of my two bags she squinched up her face in fake annoyance. "Oh, God. Am I going to have to start the day watching you drown a poor doughnut in coffee? What a miserable thing to do to a doughnut—and to a good cup of coffee, too."

Millie is a nice lady, but she is sometimes a bit more outspoken than I like.

"We all have our crosses to bear, Millie," I answered, "and yours is watching me drown doughnuts. I'll suffer splinters under my fingernails before I'll give it up. So quit trying."

"What I can't understand," said Millie, "is why, if you want to make a really disgusting mess of a meal, don't you slop some milk in the coffee, too?"

Milk in coffee is the mark of the sissy doughnut-dunker, and the look I gave her showed what I thought of the suggestion. "Even if that was a good idea, which it isn't, you never seem to have anything but sour milk for the local alley cats. So, no thanks, Millie."

She stomped from the room with a phony snarl. "One of these days I'm going to get so disgusted at the sight of you with a soggy doughnut stuffed in your mouth I'm going to get another job."

"Promises, promises," I yelled back, but she'd already slammed the door shut. I wasn't worried: Millie would never leave me. Who else but me would put up with a second mother, and a fussbudget one, to boot?

It was half an hour until the start of office hours, and there I was with coffee and two dozen doughnuts. Couldn't eat 'em *all*, of course—Millie would eat one or two at the coffee breaks, and give the rest away to the kids who would be coming in about warts, bruised knees, and booster shots—but I got first crack at 'em. I put the bag of regular doughnuts on my desk, opened the blue bag with the twisters, took one out, and swirled it into the mug.

The first dunk is the longest, to soften the doughnut, so I gave it a few seconds. Now, you expect the coffee level in your mug to go down a little, right? But would you expect the mug to almost dry up? Nope, you wouldn't, and that's why I stared at the nearly empty mug for about half a minute before I was sure of what I saw. The doughnut wasn't soggy, either, not even just a little bit damp. Just a nice, fresh, *dry* doughnut.

I took the doughnut and held it up in front of my eyes and gave it a good look. It didn't even drip. I had the same trouble I'd had at the shop—if I looked at part of the twist, the other half got sort of hazy and slipped out of view. And if I tried to focus on *that* part of the twist, I lost the first part. I shook the doughnut—no slushing sounds. The coffee wasn't in the doughnut, and it sure wasn't in the mug. *Very curious*, I recall thinking. I took the mug over to the sink in the back storeroom, filled it with water, and dunked the doughnut in it. The water level dropped from full to nearly empty in 10 s flat—but the doughnut was still dry. It didn't weigh any more than before, either, and that really puzzled me.

I snapped the doughnut in half—I noticed the hazy central part of it stopped being hazy at the instant of the snap—refilled the mug with water, and then dropped both halves in. The water level declined just a bit, and I watched as the doughnut slowly got soggy and then crumbled into pieces. *Now* it was acting like a normal doughnut. I repeated the entire business with another twister. Same results. It made the water in my mug disappear until I broke it, and then the doughnut acted normal.

I went back to my office chair, sat down, and thought what I had observed. Maybe, I reasoned, the twisters are *always* making what's next to them disappear. I could *see* the water or coffee vanish in my mug, but what about a twister I just held in my hand? Maybe it was sucking up the air, too. I held another twister to my ear, and I was pretty sure I could hear a slight rushing sound, what I imagined air being sucked into the doughnut might sound like. When I used my stethoscope, I was *positive* I could hear a wind-like sound near the center of the doughnut. I got an idea from that.

I went back to the sink and found a large glass bowl. I put a twister on the counter and plunked the bowl down over it. I waited a minute and then tried to remove the bowl. I couldn't budge it. That damn doughnut had made all the air inside the bowl disappear, and atmospheric pressure clamped the bowl down tight to the counter! I was pretty pleased at my clever experiment, until I started wondering how I was going to get my bowl back. It was a good-sized bowl, and if I broke it with a hammer the implosion would spray glass fragments all over the place. But I couldn't just leave it there, either, since the bowl might give way at any time. I went back to my desk and, after rummaging around a bit, found a pair of scissors and jammed one of its blades between the sink counter and the bowl to break the vacuum seal. It worked—with a big PLOP the air rushed back in and I retrieved my odd little doughnut.

I had another idea. I picked the doughnut up by one end and placed it on top of my desk. Taking a pencil out of the top drawer, I held it perpendicular to the desk top and pushed the eraser end into the hazy center. It should have gone in maybe an inch, and then hit the desk top. It didn't hit anything. I kept pushing until nearly half the pencil was in—and then I pulled it out. The half that had gone in didn't come out; there was just a ragged, chewed end where the eraser had been. I broke out in a cold sweat at the sight of the mangled pencil—what if I had bitten into the doughnut, or *whatever* it was? That doughnut clerk wouldn't be the only one who would need to pay a visit to a dentist!

Now I'm no dummy. I was already pretty sure I wasn't dealing with some simple doughnut salesman. I remembered how the fellow had smiled at me with a mouth like a food processor as he'd massaged my hand. I started to appreciate how an oyster must feel just before taking a plunge down somebody's throat. I sat in my chair for a long time, looking alternately at the twister and what was left of my pencil. Damn!

I decided to try one more experiment. I returned to the back storeroom and, after digging around in the closet for a bit, found Albert. Albert is a human skull I show kids while I am trying to explain to them how the earache medicine I've prescribed for them is going to help. He's a really big hit around Halloween time. Albert has a nice, firm set of teeth, and if anything awful was going to happen while eating a twister, well, better Albert than me! And anyway, old Albert had already met his maker once, and so was more prepared for whatever might happen than I was.

Back at my desk, I thrust the twister into Albert's jaws. As I closed the teeth on the hazy center of the twister there was a sudden, loud slurping sound and—Holy Cow! Albert, *all* of Albert, was sucked into that doughnut faster than a cold can of beer disappears at a baseball game on a hot day. It's funny how you do silly things in times of stress: I suppose I should have been scared witless, but all I could do was swear at that damn doughnut. Albert had been a favorite of mine.

Maybe I'm just a small-town doctor, but I do my share of reading and keeping up with the new ideas in science. It was now pretty clear to me what was going on, even if not *how*. The twisters were portals or gates to somewhere else, and it was easy to guess what happened on the "other side." Apparently, *anything* could pass through the gate, depending more or less on how easily it flowed, but it took the proximity of teeth (more likely anything with calcium) to trigger the suction into overdrive. That way, a twister wouldn't give itself away prematurely. But how could such a thing work?

I tried to remember some of the topology math I'd had in college, before I got caught-up in my pre-med classes. I recalled the old party trick of making a one-sided surface by giving a half-twist to a strip of paper before taping the ends together into a continuous loop. A Möbius strip it was called. The next higher-dimensional equivalent—called a Klein bottle—was supposed to be impossible to make in three-dimensional space. The professor had said it was necessary to give a normal bottle a half-twist through the fourth dimension, and that was obviously impossible to do.

That professor must have been wrong, though, because I'm sure that was what the doughnut twisters were—Klein bottle gateways, tasty trapdoors, so to speak, to a place I had the feeling I'd prefer not to visit. I carefully wrapped the twister in paper and dropped it back into the blue bag. I didn't want to take a chance on somebody taking a bite out of one, so I put the bag in my desk and locked it up tight. I was now getting pretty steamed-up at how I had almost been suckered, and started wondering what I could do to hit back. I thought about going back to the doughnut shop, but then what? What could I do to someone who could make a four-dimensional Klein bottle twister? I was clearly up against a very advanced technology.

I was still wondering what to do when Millie opened the door and said "We've got a slight emergency here, doctor. Buddy Hawkins had an accident a few minutes ago, on his way to school, and the nurse over there thinks he may have fractured his left wrist. She sent him over for a check-up."

"Okay, Millie," I said, relieved at the distraction, "bring him in. Put him in Room 2, and I'll be right there."

Buddy is 10 years old, and is a bright boy. I don't treat him like child, but more like an adult. He was sitting on the paper-covered examination table. He smiled at me and said "Hello, Doctor Adams." I could tell he was in pain, and there was a tear in one of his eyes.

"Let's take a look at your wrist, Buddy," I replied. "Can you flex it?"

Buddy held his left arm toward me and moved his hand up and down. His mouth pinched a little as he flexed the wrist joint, but it seemed to move freely. I gently rolled the wrist between my fingers, feeling for an obvious fracture. Then I recalled that the doughnut man had done much the same to my hand, and I stopped. "Feel any pain?" I asked.

"Not too much now. It really hurt a lot right after I fell off the tires, though."

"Well, I don't think you broke anything, Buddy. It's a pretty nasty sprain, at worst. I'll have Millie take an x-ray to be sure, but I think all we'll have to do is put an elastic bandage around it for a few days. Where did it happen?"

He gave a little shrug. "Over on the vacant lot a couple of blocks from here."

Somehow, I knew the place he meant. "The one over on Fifth and King? The one with the new doughnut shop on it?"

"Yeah," he said, "there is a bunch of tires there and some of us kids were skipping through them, like football players do in practice. Only there's no doughnut shop there. Never has been anything there but junk."

I just nodded. That the shop had disappeared as fast as it had appeared didn't really surprise me. The place probably didn't stay too long in any one location, but just long enough to unload some doughnuts and then, before people started to drop out of sight, it would move on. That clerk wouldn't actually need to catch many in his fiendish traps if he was running a gourmet shop. Humans were probably a high-priced delicacy where he came from, and he didn't require a large volume to turn a nice profit. Damn!

All the rest of the day I had to try hard to keep my mind on my patients. Always present, however, was my memory of the bag of twisters locked in my desk, and of the face of the alien shopkeeper who sold deadly traps in the shape of Klein bottle doughnuts. There seemed to be no way I could do anything about it, though, and I was pretty much resigned to the idea that I was going to have to call the police, or FBI, or CIA, and give them the remaining twisters. But that wouldn't be nearly as satisfying as doing something myself. And then I got the idea for my final experiment. Since twisters devour whatever they are near, especially anything with calcium, maybe *that* was the key on how to hit back!!

By then it was just after four in the afternoon, and I was pretty sure Jeb Starr would still be in his office. Jeb is a dentist who keeps an office next to mine. I told Millie to get him on the "phone." Jeb thought I was nuts, but 10 min later I had what I was after—four freshly pulled teeth. They had each been heavily used and abused by their previous owners, but they were just what the doctor ordered. Each was a motherlode of calcium.

I retrieved the bag of twisters from my desk and, being careful to stay away from their hazy centers, I stuck a tooth into each end of two twisters. Then I held one twister above the other (that I had laid on the desk top), with each twister making a right angle with the other. The hazy center of the twister in my hand was right over the hazy center of the twister on my desk when I let go of the twister in my hand and let it fall onto the one on the desk. It was just like throwing a bar of copper across the output lines from an electric power station.

The two twisters ate one another soundlessly, but I staggered back out of my chair when I was engulfed by an incredibly intense flash of blue-green light that illuminated the cannibalistic meal. It was a short-circuit all the way from here to there—wherever "there" was. On my desk, the bag containing the remaining twisters burst into flames, and I quickly carried the bag over to the backroom sink and ran water on it. Whatever had happened on the other side, I was willing to bet was a *lot* more dramatic than just a bright flash of light!

I dumped the bag's soggy remains into the sink, noticing with relief that none of the doughnuts had a hazy center anymore. All I had was a sink full of doughnut mush. As I looked down at that mess, I couldn't help feeling a shiver of apprehension. I knew I'd never again be able to eat a doughnut without fear. I shuddered at the thought of the kids who had eaten from the bag of regular doughnuts—what if it had been from the bag of twisters? That was a risk I could never again take. Damn!

And yet, I couldn't help smiling, too. This time the aliens had met their match. I started to chuckle at the thought of how their food system had met its end, an ending with a "surprise twist!" I felt pretty good right about then, and decided to call it a day. On my way out through the front door, I called goodnight to Millie. She came out of the small room where we keep a refrigerator for perishable items and said "Goodnight, doctor. I'll close up just as soon as I clean the refrigerator. You were right, by the way—all our milk *has* gone sour. I'll dump it down the sink in the backroom and buy some fresh milk for the cats."

I suppose I should have been quicker-witted, but it just never occurred to me that milk is loaded with calcium. And who would have imagined the aliens could have temporarily reconnected their Klein bottle link to the sink drain where I'd dumped the twisters? I was half a block down the street when an awful premonition made me turn, just in time, to see a flash of blue-green light through one of the office windows. I ran back, but it was too late. All I found were Millie's shoes on the floor by the sink, and an empty milk carton stuffed in the drain. This time the gate was finally destroyed. I guess Millie must have overloaded the link, because the sink is plugged up, too. I can only imagine what a plumber will find in the pipes. I haven't called one yet.

Poor Millie. Poor me. What am I going to tell the police?

Damn! (Fig. 9.1)

A parting comment. If this story has put you off on eating doughnuts, well, maybe that's for the best. Even the 3D versions are, you know, not good for you!

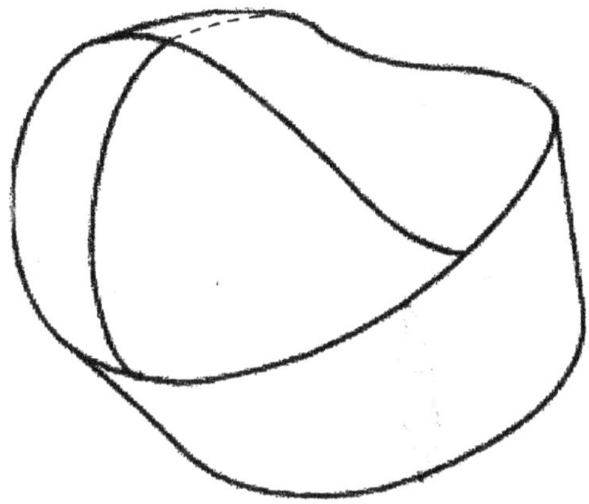

Fig. 9.1 The Möbius strip isn't nearly as optically confusing as the centers of Doctor Adams' 4D Klein bottle doughnuts are, but even the 3D strip needs to be looked at with care!

10

Alternate Universes

One of the classic derivations in college-level electrical physics is the calculation of the electric field generated in space by an *infinite plane* of electric charge, with the charge *uniformly* distributed. That is, the electric charge *density* (charge per unit area) is some non-zero, finite constant. When I first saw this problem solved (in a freshman physics course at Stanford University, more than 65 years ago), I was instantly captivated by the result—the electric field is a *constant, independent* of the distance from the plane. Two points, one just an inch from the plane and the other a light year distant, will measure the *same electric field.*

Amazing!

So, perhaps, it's understandable that years later I came to wondering what it would be like to approach an infinite, featureless plane. (*Featureless* means a perfectly smooth plane, one with no bumps or depressions in it, *anywhere*, and everywhere the same color. Think of a vastly huge, tightly stretched sheet of rubber.) For example, while thinking about the electric field calculation I wondered if you could tell, by just looking, how far you are from the plane? Thinking about that question eventually motivated "The Infinite Plane," which appeared in the April 1981 issue of *Omni*.

"The Infinite Plane" is a tale of *parallel* universes, a sub-genre (*alternate* universes includes *parallel* universes and *splitting* universes, a distinction I'll elaborate on in the next few paragraphs) that has attracted just about every science fiction writer at some point. Alternate universes occur, for example, in the 1980 novel *The Number of the Beast*, by Robert Heinlein, which has $(6^6)^6$ worlds—hence the title—that is, $\approx 1.03 \times 10^{38}$ worlds. (Heinlein could have been much more dramatic if he had written $6^{(6^6)} \approx 2.66 \times 10^{36,305}$, which

© The Author(s), under exclusive license to Springer Nature Switzerland AG 2026
P. J. Nahin, *The Art of the "What-If?"*, https://doi.org/10.1007/978-3-032-15618-1_10

requires no change in the title). H. Beam Piper (1904–1964) had even more parallel worlds in his "paratime" stories: $10^{100,000}$ worlds. So, where do all these worlds come from? It's not just imagination that offers an answer, but maybe physics, too, as I'll explain in just a moment.

The idea of multiple possible worlds long pre-dates written science fiction. It was, for example, a popular topic among eighteenth century philosophers (Voltaire and Leibniz are the usual names that come to mind). The concept caught the artistic fancy, too, of Jean-Ignace-Isidore Gérard (1803–1847) who, under the professional name of "J. J. Grandville," created the imaginative 1844 artwork shown in Fig. 10.1. His "Infinity Juggler" illustrates the idea of an *infinity* of possible worlds, which would become a popular theme in modern science fiction.

As with so much of what has become modern science fiction, H. G. Wells was there early in the fiction of parallel worlds with his 1895 novel *The Wonderful Visit*. The parallel universe idea is not central to Wells' tale, but rather is used to "explain" the background for all that comes after the first page. Definitely fantasy and definitely *not* science fiction, *The Wonderful Visit*

Fig. 10.1 "The Infinity Juggler"

is concerned with the astonishing resulting events when an angel (!) flies into our world after a Vicar fires a gun and hits the angel's wing. From just *where* the angel was, before being injured, Wells offers just a brief, tantalizing hint, one involving a fourth spatial dimension and parallel worlds with a common time dimension: "There may be any number of three-dimensional Universes packed side-by-side."

What Wells had in mind becomes clear with the understanding that *parallel* universes are imagined to be independent realities that have *always* existed, each with three spatial and one time dimension. Our world is just one of (perhaps) an infinity of parallel worlds. The time dimensions of parallel worlds are imagined to run "in sync" with each other. Science fiction since Wells has not been shy in invoking parallel worlds (see note 1 in Chap. 7 for numerous examples): typical is "The Coin Collector" (sometimes published under the variant title "The Other Wife") by Jack Finney (1911–1995). There we read of a man who has two wives, one in each of two parallel worlds between which he can travel back-and-forth at will. (This is almost certainly a fantasy for not just a few married men but, I hasten to add, not for me!) As far as I know, there is no underlying physical theory in support of the concept of parallel worlds, but that is *not* the case for the other version of alternate universes: *splitting* universes. In that case, physics does have something to say.

In 1957, Hugh Everett III (1930–1982) submitted a doctoral dissertation to the Princeton University physics department on the foundations of quantum mechanics ("The Theory of the Universal Wave Function"). Everett's idea was later given the name by which it is known today ("The Many-Worlds Interpretation of Quantum Mechanics") by the physicist Bryce DeWitt (1923–2004), often called the MWI. The MWI makes what strikes many to be a fantastic assertion: with each quantum-level decision (for example, the "decision" made, at each instant of time, by a radioactive atom on whether to decay or not) the universe splits into branches of reality, with each equally real branch associated with one of the possibilities for the decision.[1] The process than repeats, endlessly, for each of the branches, with each branch spawning yet more branches. The continuous splitting of reality results in a multitude of worlds beyond counting, a multitude that becomes immensely bigger with every passing instant.

It's easy to see why many physicists think the MWI to be absurd, and critical commentary on it has included the charge it is "bizarre," and that it is "not easy to reconcile with common sense," and that "if such a theory were taken

[1] At the very end of Chap. 16, I offer a fictional theological mechanism for how quantum-level decisions are made—but I urge you not to peek just yet!

seriously it would hardly be possible to take anything else seriously." Everett's own doctoral advisor, the highly respected John Wheeler (1911–2008) who of course signed-off on Everett's dissertation, eventually disavowed it, saying (in a *huge* understatement) that "its infinitely many unobservable worlds makes a heavy load of metaphysical baggage"—see the following box for a theological puzzle that the MWI presents that is probably *not* what Wheeler had in mind, but about which, I suspect, he would have had something interesting to say.

When the Universe Splits, What Happens to God and Satan?

The idea of reality splitting has an interesting metaphysical twist to it. Suppose God is unique, an assumption made, I believe, by most religions (at least implicitly). That means God does *not* split at each quantum-level decision, and so He must choose, at each splitting, to follow just *one* of the branches of reality that results from the split (and so *not* to choose to be in any of the other realities). On those not-chosen branches there is no God. Given how many times reality has split since the Big Bang, the probability is virtual certainty that God is not present on an arbitrarily selected reality—like ours? And, of course, this same argument can be made for Satan. So, there must be *one* reality with God *and* Satan (is this our world?), exactly *one* reality with God but no Satan (*that's great*!), exactly *one* reality with Satan but no God (*that's not great*!), and all the rest of the realities having both God and Satan *absent* (*so in those realities there's no Heaven or Hell and you're on your own*!).

There must be a good science fiction story in all of that, somewhere!

Many science fiction writers, unlike most physicists, *love* the MWI for its ability to "explain" time travel paradoxes (or at least some of them). The idea is that if a time traveler journeys into the past, and so introduces a change, then reality splits into two branches, with one branch representing the change and the other reality being the original reality without the change. What really puffs these writers up with pride is that the splitting universe idea *appeared in the science fiction pulps decades before Everett's Princeton dissertation*: in the story "The Branches of Time," in the August 1935 issue of *Wonder Stories*, written by David Daniels (1915–1936).

"The Infinite Plane" has no time travelers, and it has just two parallel worlds. I agree with those critics of the MWI that it is a "bloated" theory, and so when I set out to write an alternative universe story, I purposely avoided the MWI. In my story, I attempt to offer an alternative "explanation" for the two parallel worlds, but of course it's all just (*I think*) made-up. As you read "The Infinite Plane," keep an eye out for the "explanation."[2]

"The Infinite Plane"

The North Korean MiG-15 swept up from below and behind on a negative-gee attack run, its 23-mm and 37-mm cannons flickering red like the tongue of a poisonous snake. The intercom circuits crackled with the tail gunner's shout of "Single bandit, six o'clock low! He's hot, oh baby, he's hot!" The enemy fighter arced smoothly out of its climb—its pilot was very good—and a stream of red-hot shells, each a demon's fireball as big as a man's fist, stitched through the tiny world of Sergeant Richard Mackley and cut him free from the B-17.

He began to fall, plummeting toward the Sea of Japan, 25,000 feet below.

He had been tucked into the underbelly ball turret, his knees pushed up into his chest, with his eyes pressed to the gunsights. His hands had been above his head on the position controls, and his feet flat on the gun pedals. He had been trying, *very* hard, to kill the enemy pilot with the turret's twin 50-caliber machine guns.

The Sperry turret had twisted him around and up and down at a dizzy rate as he'd tried to direct the double stream of bullets into the MiG's canopy. Just like he used to shoot the garden hose at butterflies while watering the lawn when he was a kid. He'd been pretty good at washing butterflies out of the sky, but that had been a long time ago, and he was 22 now. That had been a child's sport. This was war, and death, and for real.

And this time he lost. This time *he* was the butterfly, washed clear of his bomber and falling through the crystal-pure, 30-degrees-below-zero air. Only his flight suit, and both lungs full of one last, deep, lucky gulp of pressurized oxygen, saved his body and mind from instant destruction. Richard Mackley's first thoughts as he fell away from the machine that had protected him from gravity were remarkably concise, and calm. *Shit! That bastard sure got lucky. One more second, just one, and I would've put some rounds in 'im.*

His lack of fear surprised him. He had always been sure that, if faced with a violent, certain death, he would scream. And he was surely about to die unpleasantly since the turret was far too small for its gunner to wear a

[2] In Chap. 14, I set a story ("Monster's Meal") in a parallel universe that can be visited with the aid of a crazy gadget called the "Hyperspatial Mega-Time Warper." In "The Infinite Plane," I tried for a somewhat more plausible physical explanation!

parachute. Maybe his calm was because he'd long ago resigned himself to the belief he'd never live through the war anyway. What did bother him was the foolishness of it all. He shouldn't even be here. It wasn't the enemy pilot's fault. The whole mission, itself, had been a wild-goose chase. It was the fault of that cocky, high-strutting, steel-rod-up-the-ass intelligence major! Listening to him at the preflight briefing that morning had nearly made Mackley puke.

The major was glorious in his freshly pressed uniform and Clark Gable mustache. He was pointing at a map of Japan and Korea with a thin metal rod that projected a small, bright spot of light. Mackley looked around the room and saw just his crew and the major. A relaxed, neatly turned-out gentleman officer and ten war-weary airmen wondering just what the hell was going on.

"All right, men, let's get to it. I know you must be wondering about this surprise mission. We've just learned the Red Chinese have moved a large number of men, possibly of divisional strength, across the Yalu River and into Chosan. We *think* they may be getting ready for a big thrust into the Iron Triangle, here, a bit west of Kumsong." The spot of light held steady on the map, illuminating an area just above the 38th parallel. "Division HQ wants a photo-recon to back up their recommendations to the brass in Tokyo. That's where you fellows come in."

Captain Lynn had stuck his hand up then. Sergeant Mackley liked the captain, respected his common sense. He was a tall, almost emaciated man, but with a deep, booming voice that sounded out of place coming from a throat where the Adam's apple bobbled in plain sight. "Why us, Major? Why a slow-moving B-17 that's already had more than her share of grief, when a Sabre jet with a hot jockey could roar in there and out a lot faster than we can?" Yeah, Mackley sure liked the captain. *That* was the question in his mind, too. Why, indeed, did they have to fly baby-bare-assed into enemy air-space, lumbering along at less than 300 miles per hour, just daring speed-smoking, cannon-firing MiGs to come up and shoot their brains out?

The major looked peeved. He'd had to deal with these damn combat types before, and it pained him to have to explain why things were being done as they were. Actually, it was none of this pilot's goddamn business *why*, but the major answered anyway, although his expression showed he considered the captain's question out of line. The major held up a hand and began to flick fingers up, one at a time.

"There are several reasons, Captain. Let me go through them, one-by-one, for you."

Flick

"First, it will take too long to fit a camera mount to an available F-86 Sabre, and HQ wants these pictures ASAP."

Flick

"Second, the Commies maybe won't think much of a lone, low-speed radar contact, while a fast-moving fighter is much more likely to provoke a response."

Flick.

"Third, if you do get a reception, a heavily armed bomber can keep right on boring in, able to defend itself with its guns without taking escape maneuvers that'd ruin the photo-run over the target."

Mackley felt the sweat running down his back. *Lone radar contact.* That meant no fighter cover. He saw Captain Lynn stare hard at the major, but say nothing. The captain knew it would do no good, would only cause trouble. Mackley understood that, too, and felt pride at serving under such a cool-headed leader. If anybody could get them in and out in one piece, it would be the captain. And the bomber was a lucky machine. She'd been shot-up bad before, but she'd always come back. Captain Lynn had said it had flown over Berlin, just before the end of the last war, when the German air defense had been desperate, and she had survived. Maybe that luck would bring her and her crew back from this mission, too.

Mackley's thoughts came back to the present, and he realized he was tumbling. He couldn't actually see the sky and water rotating about him because the cloudless sky seemed to be a blue that matched the color of the ocean. It was the same blue all about him. Still, he knew he was tumbling because there was a strong centripetal force and he was pressed flat against his back onto the cracked Plexiglass of the turret. The sudden realization of the ruptured nature of the turret housing made him wonder why he was still alive. He should have passed out from lack of oxygen, since his breathing hose had fed him from a pressure tank in the bomber. And that tank was a long way off, connected to the other half of the severed hose.

The bomber had to be a long way off by now, didn't it? He'd been falling for—how long? Mackley tried to figure how long it had taken him to remember the morning events, but he couldn't. It must have been at least 30 s, though. How far could you fall in 30 s? *Let's see. What's the formula from physics?* Mackley had been a good science student in high school and the first 2 years of college he'd finished before dropping out to join the service. Then he had it.

Falling from rest, in a vacuum, it was acceleration multiplied by the time squared. Divided by two. Forget air drag for now. *Jesus Christ! I can't believe I'm so calm!* So that'd be 32 feet per second per second, times 30 s squared. Divided by 2. And that'd be, say, thirty times nine hundred, or 27,000 feet. Divided by 2. He'd fallen over 13,000 feet, over two miles. That was half the way down! He'd wasted more than half of the precious seconds left to him thinking about that goddamn major! His sudden anger made him panic.

He had to get out of the turret!

Mackley struggled to turn around, so he could thrust himself headfirst out of the hole in the turret. There wasn't enough room to do that. So he pressed himself, rump first, through the busted opening. For a confusing instant his pistol holster tangled on something and then broke free. Mackley wished he'd worn a parachute instead. Just before he popped himself out he remembered to put his flight goggles on, to keep the wind blast from punching his eyeballs back into his brain. He was still a dead man, but he had to get out of that turret!

As he pushed free of the wrecked sphere, Mackley was surprised to see it fall away from him. He felt the slip stream alternately blast into his face and his back as he tumbled, but when it slapped him in the face he knew was looking down. And then the turret was far below him, and it was moving even farther away. The turret was falling faster than he was, quietly fading away out of sight, into the universal blue that engulfed him. Then he remembered air drag, and terminal velocity.

Terminal velocity—the speed at which the force of gravity pulling him down was exactly counterbalanced by the air drag. The still fairly round, aerodynamically shaped turret had a terminal velocity greater than his flapping, sprawling body. That meant he hadn't fallen as far as he'd figured. He had more time left. It was crazy, but the temporary stay of his death calmed Mackley, soothed him. *What's the terminal velocity of a human body?* Mackley had heard it could be as low as a hundred-twenty miles per hour, say, a hundred eighty, two hundred feet per second. *If it was spread out flat.* That'd give him—a quick calculation—maybe another minute or so until his body splattered into a big, thin pancake. *What a crappy way that would be to die!*

Maybe, Mackley thought, *if I make myself into a vertical shaft, I can knife right into the water!* He laughed at the idea—he'd heard that suicides off the Golden Gate Bridge in San Francisco, a mere few hundred feet up, were often found with their hearts in their lower gut. Or in their throat. It all depended on whether they hit feet-first or head-first. The body decelerated fast when it smashed into the water, but the soft internal organs obeyed the law of inertia faithfully and kept right on going, ripping free of their supporting tissue.

Would his brain turn to puree and squirt out of his ears? Would his eyes pop out of his head like marbles?

He remembered the 45 pistol, strapped to his waist. *Maybe I can slow myself with the recoil by shooting downward—yeah, if only it were a howitzer!* Mackley laughed, and wondered whether he was losing his mind.

He laughed again, now hysterically, as he recalled a childhood nightmare, falling free in an elevator car that had broken its cables. For a long time he'd wondered whether you could survive such a fall by timing the impact so perfectly that you jumped upward at the instant of collision. *Maybe I can jump up just as I hit the water.* He started to scream at the insanity of that, and kept screaming until he had to stop to suck in more air. Mackley screamed for a long time.

And then a death-defying thought: *I should have hit by now! Christ, I'm still alive! What's happening?*

He was falling facedown, eyes open, the howling wind blowing his hair straight back, his body all spread out. All prepared to set the world record in the high-dive belly flop. Mackley twisted his head around, peering, straining his vision through the flight goggles into the pure-blue nothing that was all about him. *Nothing, no waves, no clouds, no horizon. There's nothin' down there! Goddamn it, there's no water below me. I should see waves or whitecaps by now. There's nothin' but that damn blue. Where the hell am I?* Understanding came to him slowly, and he fought against it, unwilling to dare to hope that he might really have a chance, that he just might actually live! But there was no other explanation, and it *had* to be the answer! He scanned his memory for the long-forgotten idea: it *had* to be the answer.

Richard Mackley was falling toward Professor Hightower's infinite plane!

Professor Hightower was, as he lectured, in his customary position before his freshman mathematics class, leaning against the blackboard with his hands and pants and coat covered with chalk dust. His neatly trimmed beard was the same color, red, as his glass frames, making the glasses look as though they grew right out of his head.

"Yes, ladies and gentlemen, the infinite plane is not generally appreciated for its fantastic properties. You all know what a mathematical plane is, of course, from high school geometry. It's a flat surface with zero thickness. The sort of stuff they make desk tops out of." The class laughed at that, even though it wasn't really all that funny, but he was one math professor who at least *tried* to keep his students interested. "But a desk top isn't an *infinite* plane. It is merely a segment, an isolated part of the amazing whole. Nobody has ever actually seen an infinite plane, of course, but nevertheless we can imagine it. That is the beauty of mathematics and the human mind, you see. *Together*, they can create what Nature herself has, for some reason, seen fit not to make."

"And why would Nature make such a sad decision? Ah, but you must remember that *infinite* means just that. Not big, not huge, not spectacularly enormous, but *infinite without bound*. It would stretch in every direction from one end of the universe, to all the way across and beyond. It would divide the universe into two separate parts, and nothing on one side could go around this majestic membrane to reach the other side."

The professor pointed at Mackley who was sitting in the front row. "Tell us, young man, why is that so?" Caught momentarily off guard, Mackley nonetheless quickly answered, as he found the idea of the infinite plane strangely fascinating.

"Because, Professor, since it is infinite, *it has no edge to go around!*"

"Correct! The infinite plane would, in effect, create *two* universes where there had previously been just one. Each universe would be isolated from the other, and the only way to leave one universe and enter the other would be to punch *through* the infinite plane, to rip a hole in it. But that, alas, would ruin the infinite plane's perfect beauty."

The students chuckled again at the professor's mock grimace of horror at the thought of doing such a terrible thing.

Professor Hightower leaned forward from the blackboard, eyes now gleaming, and spoke in a hushed tone. "But I haven't yet told you the most incredible part. You all know how objects change their appearance as their distance from us varies. To be completely trivial about it, the closer something is, the bigger it looks. Now, let me ask you the penultimate question for today. Suppose you were, somehow, suspended a thousand miles above such a plane. How would its appearance change if you were dropped to a distance above it of a mere one mile?"

The class sat stunned. Boy, the professor had really flipped out this time! The old geezer must be nuts. How could you possibly figure out how something that didn't even exist would look? Only Mackley thought seriously about the question and, finally, he raised his hand. The professor seemed both surprised and pleased to have a response.

"Yes, yes, young man, let's hear from you again."

"Well, Professor, I'm probably wrong, but it seems to me that if it's infinite in extent, and if it's perfectly flat, then—well, it would look the same, no matter how far away or close to you it actually was."

"Correct, once again! Very good, excellent. How, indeed, could it look different? It couldn't look larger, just because it's closer, since it *already* is infinite. There is no edge to the thing, so there'd be no horizon, *nothing*, to serve as a depth cue for your eye. Therefore, it would look the same. QED, as we mathematicians like to say."

"Now, as I said, that was the penultimate question. So here's the last one, a *hard* one, maybe, to answer—and just in time, too, as the hour is nearly gone. If you could, that is if you really were on one side of an infinite plane, *would* you have the will, the sheer, raw nerve, to rip through it to see what was on the other side? Think carefully about that one, ladies and gentlemen. Think carefully!"

"Remember, closed-book exam on Monday. Class dismissed!"

It all fit together so perfectly. It was an infinite plane, not the hard ocean, that was below him. But how had he gotten here? He'd been in a heavy bomber, his body jerking to the booming vibrations of the turret's machine guns, and now he was—*where*? As the professor had said, he could be a thousand miles above it or maybe just a thousand feet, There'd be no way to tell until he punched through it. Mackley had heard another professor, in an introductory physics class, talk once about alternate universes, places outside the one they lived in. Maybe that was where he was. But how had he gotten here? He struggled with that question, yearning to know the answer to what had saved his life.

Then he recalled part of an overheard conversation between two Air Force scientists when he had been in gunnery school. There had been a research symposium at the training base and once, while on an errand to the auditorium building, he'd walked for a minute or so down a long corridor behind the two men. They'd been working on the problem of measuring the trajectories of high-velocity bullets, trying to find ways to improve the lethality of airborne guns. They'd talked about how fast-moving bullets create an intense electrical plasma of hot gases around themselves as they rip through the air.

So maybe *that* was the answer. Maybe those MiG cannon shells had those plasma clouds, too, and when the shells tore the turret loose their interacting electrical fields had been just right, somehow, to rip him from the ordinary universe and to transport him to this goddamn torture of a fall that just never was going to end and—Mackley's heart skipped a beat as he pursued that thought. *Jesus, what if this fall really never does end? What if I just bounce off the infinite plane? What if it doesn't rip? Maybe I'm already dead and gone to Hell? Maybe that's my hell—to bounce forever on the infinite plane that's the Devil's goddamn trampoline!*

Mackley gave up. *To hell with it!* The wind blast was still tearing at his clothes, the blue was still all around, he was still alive and so to hell with it. He was going to make it! Mackley felt exhilaration sweep through him. He was tired of the damn war: it *had* been getting boring, just waiting to get blown away. Here was something new and exciting, and he was at last free of the war. What *would* be on the other side of the infinite plane? Wouldn't old

Professor Hightower give *plenty* to be here with him! Mackley then remembered a hallway conversation he'd had with the professor a few days after the infinite plane lecture. He'd been walking down a corridor on his way to another class when Professor Hightower had come out of a nearby classroom. The professor had a sharp eye for faces.

"Hello, young man! Mr. Mackley, isn't it? I want to congratulate you on the fine exam paper you wrote last Monday. Excellent piece of work, and I could see from it that you'd been thinking about what we discussed in class concerning the infinite plane. I wish all my students were as diligent as you."

Mackley had blushed with both pride and embarrassment. The professor, a naturally exuberant man, had spoken loudly, and other students in the hall, hearing his praise, were staring at them. Startled, Mackley replied "Thank you, Professor, but it wasn't anything special. It's just that the infinite plane is so interesting." The professor had smiled and Mackley felt encouraged to continue. "I've been wondering. What do *you* think would be on the other side of the infinite plane?"

Professor Hightower had laughed. "Oh, my dear young man, *that is* the question, isn't it! My speculation is no more profound than anyone else's but, I do, of course, have my own theory about it. Nature, you see, loves simplicity and symmetry. It is Occam's Razor, the law of parsimony, that says it best. The simplest possible explanation of nature is the true explanation." Mackley must have looked confused because the professor had smiled again and said "You are taking physics, too, aren't you?" As Mackley nodded, the professor had continued with "Yes, of course you are, and it was that great physicist Isaac Newton himself who said it so well: *Nature does nothing in vain, and more is vain when less will serve. For Nature is pleased with simplicity and affects not the pomp of superfluous causes. Thus we, too, are to admit no more causes of natural things than such are both true and sufficient to explain their appearances.*"

Mackley had still been confused by all that, and the professor had sounded a bit exasperated with him. "Well, don't you see it yet, young man? The simplest thing for Nature to do would be to have one side be the image of the other, an exact reflection. The infinite plane would separate two duplicate universes. To penetrate the infinite plane would be like somehow passing through the surface of a mirror, like Lewis Carroll's looking glass, and entering a replicated world."

Mackley didn't like that theory. To have gone through all of this just to fall into a universe identical to the one he'd been in—that wouldn't be fair. *There has to be more to all this than that. There just has to be.* There might, maybe, be a primitive world on the other side, a world with people who would think his advanced technical knowledge to be magic. He would be a god in such a world. Or maybe it would be a universe where all the physical laws are different, where everything would be wonderfully strange. Mackley wasn't sure he'd like that kind of world. But of course it might be nothing like that at all. Professor Hightower had been right about the infinite plane's existence, so maybe he was right about the mirror-image universe, too. Mackley couldn't wait to tear through the blue barrier and see what was really there. He spread-eagled himself as far as he could, arms held at right angles to his body, legs extended stiffly. He stared straight down into the deep, featureless blue that seemed to go on forever and yelled "Here I come, you beautiful infinite plane! Here I come, and oh, man, am I going to blast right—".

Mackley hit the membrane, and time seemed to slow to a crawl. It was only now, as he pushed into the blue, that he could see how slowly things were happening. It was like falling through molasses. *Could that be why the fall went on for so long? Is the time flow different here? But if that's so, why aren't my thoughts slower, too?* Mackley then realized that if he was thinking more slowly there would be no way to perceive the slowness. *Am I going mad?*

He could feel the infinite plane's resistance to his slowly hurtling body, sense its stretching, see the depression he was making in it. Now that he was actually in contact with the membrane, it *did* have a texture, a cool, smooth surface to it, and he could see the flatness of it curve inward beneath his body. And yet, somehow, he knew it wasn't a material object. It was a force-fabric woven from the fibers of space-time. He caressed it lovingly with his hands and legs.

And nearby, only a few meters away, he could see a mirror-image bulge forming from the other side. As he slowly sank inward, the other bulge, equally slowly, pushed outward. Deeper, deeper, he penetrated in one direction, and further, further outward the mirror-image bulge grew. Something was trying to break through from the other side. That meant there *was* another world there, that the professor had been right. The infinite plane, and the region of space-time around it, was the interface between two universes, a gateway between two halves of reality.

He ripped through—and there was just a glimpse, a mere flash in the corner of his eye, as the mirror-image bubble ruptured simultaneously. The shock made him yell, but the sound was lost by the muffling of the infinite plane.

It was himself, his otherworld twin!

And then it, or he, or *something*, was gone, whipping away into Richard Mackley's lost world just as he was falling into the new one. Now he was rushing through the air again, and it was like a film running backward at high speed. The wind blast tore at him from above as he streaked upward, and the surprise of it all left him stunned.

The intercom circuits buzzed through his mental haze, and the voice of Captain Lynn jolted him back to alertness. It wasn't possible, but he was back in the turret, back in the bomber! "Mackley, Mackley, you there, boy? Come on, answer!"

"Mackley here, sir. I'm sorry, I must have gotten disoriented with that last firing pass."

"Well, snap to it, Mackley! Goddamn it, boy, this ain't no time to crap out."

Fortunately there were no more MiG attack runs as Mackley collapsed against his turret controls, almost in a faint. *Jesus, I dreamed the whole damn thing! I gotta see the doc when we get back. I must be cracking up, heading for a section eight and the psycho ward!*

After the bomber landed safely back in Japan, Mackley and his crewmates walked to the intelligence shack for the post-flight debriefing, but he avoided joining in with the tension-releasing banter of the others. His legs were still wobbly, and he was on the verge of throwing up. In the shack, they took pencils and paper from the waiting clerk and began to fill-in the standard post-mission question sheet. Halfway through, Mackley suddenly froze as understanding hit him, numbing him with fear. He *hadn't* dreamed it, it *hadn't* been a hallucination. *It had all been real.* He had misinterpreted Professor Hightower's explanation of a looking-glass world. Reflected worlds would *not* be the same! Everything looked normal on the surface, but he knew that he was in an alien place. Was it just his imagination? Or were the others glancing strangely at him? Could they sense that he was different, and that he knew he was different?

Mackley stared now in horror at his pencil, held tightly in his left hand. That was simple enough, wasn't it? But it told him that things were never going to be simple again. When Mackley had begun the mission, he had been right-handed.

The idea of reversing the "handedness" of an object has a long history in science fiction tales that involve the fourth dimension. That idea goes back to 1827, when Möbius (remember him from the last chapter?) showed that a 3D object is converted to its mirror image if flipped over through the fourth dimension. H. G. Wells used that in his "The Plattner Story" (in the April 1896 issue of *New Review*), which is the tale of an experiment gone terribly wrong: an explosion blows a chemistry teacher (Gottfried Plattner) into the fourth dimension and, when he eventually returns to "our world," it is found that he has been "flipped." His heart is on the right-hand side of his chest, he is now left-handed when he was originally right-handed, and so on.

The story provides a brief explanation (particularly important to its original Victorian-age readers) on what has happened to Plattner as follows (remember, this is *decades* before the appearance of the science fiction pulps):

> There is no way of taking a man and moving him about in *space* [Wells' emphasis, in a reference to 3-D space], that will result in our changing his sides. Whatever you do, his right is still right, his left his left. You can do that with a perfectly thin and flat thing, of course. If you were to cut a figure out of paper, any figure with a right and left side, you could change its sides simply by lifting it up and turning it over.
>
> But with a solid it is different.[3]

Fifty years later Arthur C. Clarke gave Wells' story a modern science fiction treatment in his tale "Technical Error," published in the December 1946 issue of the short-lived (only three issues) British science fiction magazine *Fantasy*. There we read of an electrical engineer who is reversed when caught-up in an electrical short-circuit of enormous magnitude in a power station, a short-circuit which results (we are told) in rotating a volume of 4D space (with the engineer in it). When reprinted in the June 1950 issue of *Thrilling Wonder Stories*, the title was changed to the more descriptive "The Reversed Man,"

I decided, when writing "The Infinite Plane," to avoid any mention of the fourth dimension, but rather to leave it up to readers accept the "physics" of penetrating an infinite plane as being behind Mackley's reversal.

[3] Wells' last two sentences are important. Wells' example of "a perfectly thin and flat thing" is of a 2D object flipped over in its *hyperspace* of 3D space. To so flip a 3D object you need access to *its* hyperspace (which is 4D). Wells continues his explanation with "Mathematical theorists tell us that the only way in which the right and left sides of a solid body can be changed is by taking that body clear out of space as we know it [and] turning it [over] somewhere outside [our 3D] space."

11

National Security

When I was in grammar school in the late 1940s, and in high school in the mid-1950s, children routinely worried about dying in an atomic war. Those were the days when kids were taught how, at the sudden appearance in the sky of an intense flash of light, to throw themselves to the ground and to cover their heads (or to throw themselves under their desks if in a classroom). Everybody knew what "duck and cover" meant, and the threat of nuclear catastrophe always lurked in the consciousness. That was a pretty terrible situation, but of course with time has come new terrors: today's kids worry about dying at the hands of a mass murderer armed with a military-style combat assault rifle designed specifically for war, a murderer who has just shot their teacher dead and who is now aiming at them.

When I wrote "Security Blanket" for the March 1981 issue of *Analog Science Fiction Magazine*, the idea of an entire classroom of children being slaughtered in a hail of bullets would have been dismissed as simply unthinkable. The Columbine High School tragedy, which seems to have been the first of the now periodic mass shootings in America, was still nearly two decades in the future when I wrote the story. Not being able to imagine such a thing myself, "Security Blanket" is centered on the situation I remembered from my own school days. The "solution" proposed in the story, I regret to say, would be of no use for today's children, and so perhaps in that sense "Security Blanket" is not a success. However, it does offer, I think, a look back at a time when the threat of atomic war might be considered a *lesser* one than is the modern horror.

And that is really pretty sad.

© The Author(s), under exclusive license to Springer Nature Switzerland AG 2026

P. J. Nahin, *The Art of the "What-If?"*, https://doi.org/10.1007/978-3-032-15618-1_11

"Security Blanket"

The nuclear cruiser USS *Long Beach* was in the fifth day of her mission of beating the Barents Sea into a beer-like froth. Running at her cruising speed of 18 knots at 40° longitude East of Greenwich, and at a latitude about that of Smidovich, she towed the enormous paddle-wheel contraption behind her in an embarrassed silence. A silence broken only by the rhythmic slap-slap-slap of 30 ridiculously huge, 10-m-wide blades.

The crew of the Soviet[1] surveillance submarine tailing the *Long Beach* couldn't understand what the crazy Americans were up to. The Americans on the *Long Beach* didn't know either. They were just following orders. Told to tow the spinning monster, they towed it.

The Russians, using a narrow-beam, anti-intercept satellite channel, transmitted their observations back to the Committee for State Security.[2]

It was Bobby Wilson's mother who first noticed that her son was strangely quiet. It wasn't normal for any 9-year-old boy to just sit in a chair, staring out the window for 20 min without moving. It certainly wasn't normal for Bobby Wilson. He had just gotten home from school, and Mrs. Wilson wondered if he might not have picked up the flu bug from one of the other children. She knelt beside him and pressed a hand to his forehead.

"What's the matter with you, Bobby? Here, let me feel your head for a fever."

Bobby didn't try to twist away like he usually did when his mother got worried about his health. He just mumbled "Ah, mom, I'm not sick. I'm just thinking, that's all."

His mother stood up and looked at him, her hands on her hips and a worried look on her face. "You sure? No stomach cramps, or sore throat, or maybe a headache?"

Bobby shook his head no, but said nothing. Now that *really* worried Mrs. Wilson—she liked to joke that the only way to shut her son up was with a big cork and a hammer. And then Bobby did something that shocked Mrs. Wilson, and made her turn away without another word. She decided right then and there that whatever was bothering Bobby was serious enough to discuss with her husband.

---

[1] The Soviet Union collapsed (as I mentioned in the story at the end of Chap. 2) in 1991. If writing this story today, I would call the sub a *Russian* sub.

[2] The Committee was known in the West as the KGB when I wrote this story (in Romanized Russian, the name has the initials KGB). It survives today (under a different set of initials) doing exactly what the KGB did.

"He was *sucking his thumb*? Well, kids sometimes *do* that, honey, it's nothing to get all worked up over." Mr. Wilson was trying to read the evening paper before dinner, and still not get into trouble with his wife by obviously ignoring her. It was a hard act to pull off, and 20 years of marriage hadn't dulled the challenge. This time, though, it didn't work.

"Tom, this is serious. Now *listen* to me! You know Bobby would rather hit his thumb with a brick than get caught sucking it—not after all the teasing he got in school over it, He hasn't done it for over a year, and now he's back doing it. His behavior is regressive. He's obviously under some sort of stress, what with his moping about the house. He won't tell me a thing, but maybe you can find out what's troubling him."

Tom Wilson could see it was no use. Putting his paper down with a sigh, he replied "All right, Mary, where is he? In his room?"

"Yes, he's been up there all afternoon. Why, when Peter White, Bobby's very best friend, came by a couple of hours ago to ask him to go frog-catching over at Jackson's Pond, Bobby said *no*. That *proves* something is wrong, Tom!"

In spite of himself Mr. Wilson couldn't keep from grinning. "Begged off on frog-catching, huh. By George, Mary, that's proof enough for me. That's *definitely* not *my* boy! I'll go have a little talk with him." Tom Wilson put his paper aside and went upstairs to his son's room. The door was open, and he could see Bobby sitting at his desk. The room was in a general state of chaos, but that didn't concern Bobby's father. Mr. Wilson's lack of neatness caused his wife nearly as much despair as did her son's.

"Got a minute to talk, son? Your mother is sorta worried about you, and I thought maybe we could talk it over together." Mr. Wilson could see his wife had been right. Bobby *did* look unnaturally serious.

"Sure, Dad. I'm sorry if I got Mom upset. But I'm not sick. I've just been thinking, that's all."

"That's good, Bobby, but whatever you're thinking about must be pretty serious to keep you from frog-catching. Wanna tell me about it?"

"Yeah, I guess I *was* kind of rude to Petey. But I'll go to the Pond with him tomorrow. I promise. I just didn't feel like it today."

Mr. Wilson carefully picked his way through the model airplane parts and the stray clothing on the floor and sat down on Bobby's bed. "Okay, sport, shoot. What's on your mind?"

Bobby swung his chair around to face his father, and looked him straight in the eyes. "Dad, could the *whole world* blow up? Could *everybody* be killed?"

Now *that* startled Mr. Wilson, but he could see Bobby was very serious. "What makes you ask that, son? That's a pretty heavy question for a champion frog-catcher!"

"It was in science class today, Dad. Mr. Harris was talkin' about energy and stuff, and how the world is running out of oil. And he talked about atomic energy and how all the countries in the world are building bombs and other horrible things, and how they were so powerful they could kill *everybody*! Is that true, Dad?"

Mr. Wilson thought carefully before answering. *Boy, the world sure has changed since my father was a kid. What a damn shame children have to grow up worrying about such nightmares.*

"Well, yes, son, Mr. Harris *is* right. Atomic energy *does* have the capability to do harm. But *only* if people aren't wise enough to use it carefully. It's just like fire, Bobby. Fire is a wonderful thing, *if* you use it wisely. It warms us in the winter, and it can cook our food, and it can do a lot of other amazing things, too. Like make steel for cars and ships, and buildings. It can also burn things, and people, too. It must be used wisely. It's the same with atomic energy. It can be used for good *and* bad, but it's *people* that decide that."

"Yeah, that's what Mr. Harris said, too. But he also said that atomic bombs are so big that even just one could kill maybe *millions* of people. And that if we had an atomic war that the whole Earth could be destroyed! Is that right, Dad?"

"Yes, it is. But you should also realize, Bobby, that such a war isn't very likely. Atomic weapons have been around for a long time, and except for just two times, they've never been used. And even then, there was a warning given, and we were already at war."

Bobby looked unconvinced. "I was thinking, Dad, since what Mr. Harris said, about how everything would be smashed if there was a war like that. You and Mom, and Petey, and the frogs in the Pond—and me—we'd all be dead! I'm afraid, Dad!"

Mr. Wilson could see tears in his son's eyes, and it made him bitter. Why *should* a little boy have to have such a bloody sword of doom hanging over his head? Why couldn't kids enjoy their first years like kids used to, believing they were immortal? *Nowadays a kid gets to be old enough to understand words and he's immediately overwhelmed by the possibility of his own death!* It was that thought that made Mr. Wilson tell his little white lie. He ordinarily believed in being totally honest with his son, and Mary agreed with that—"if he's old enough to ask, he's old enough to be told."

But this wasn't about the birds and the bees—the poor kid was worrying himself sick about dying when he shouldn't be concerned about anything more complex than catching a frog. When Bobby got older, and was more mature, that would be the right time for him to worry about the difficult world he lived in. But not now; it was too soon!

"Bobby, I'm going to tell you something, but I want you to promise you'll keep it secret. Just between you and me. Okay?"

Bobby wiped his eyes with a casual swipe of a sleeve (he didn't want his father to know he'd been about to cry) and said "Sure, Dad, I'm real good at keeping secrets."

"I know you are, son. Well, you know I'm not a scientist, but I do work at the government lab in town. So, I hear things about what the scientists there are doing." He leaned forward and dropped his voice to a conspiratorial whisper. "I hear there's a new electronic gadget the government has developed that can protect against atomic bombs. It makes a sort of invisible bubble over a city that you can see through, walk through, even fly through. But it stops atomic bombs. How it works is a big secret, of course, and I don't know anything about that, but even the fact it *exists* is a secret, too. So, you shouldn't talk about it. Anyway, don't you worry about any atomic war!"

Bobby looked relieved. "You mean—maybe one of those bubbles might be over *us*, right now?"

"I can't say *for sure*, Bobby, but like I said, from what I hear, it seems like maybe that's the case."

Later that same evening, after their son was sound asleep, Mrs. Wilson asked what had happened to bring Bobby back to his normal, cheerful self. Mr. Wilson told her of the fib, and while she at first worried over the correctness of telling such a tale, she soon agreed the result had been good. "After all, dear," Mr. Wilson had said, "*everybody* needs a security blanket of one sort or another."

Nearly 3000 miles to the west, at about the same time the Wilsons were discussing the right and wrong of lying to children, 2 USAF helicopter pilots started their regular late-afternoon mission of pushing a helium balloon over the California countryside. Flying at 500 feet, and tipping their huge machines backward at a dangerous angle, the blade-wash shoved the enormous, 80-m in diameter, silver-glazed sphere along at about 5 knots. The flight plan was different each day, and on this particular one they were to push the balloon back-and-forth along a line joining the Marine Naval Air Station at El Toro, and March AFB in Riverside. They kept it up for several hours, thus ensuring that the Soviet high-resolution spy-in-the-sky photo satellite orbiting overhead would get a large number of superb pictures.

The local citizens had a wonderful time trying to figure out what the Air Force was up to. They had no more success than did the KGB photo-interpreters in the basement of Moscow's infamous Lubyanka Building at 2 Dzerzhinsky Square, who also puzzled over what the crazy Americans were doing. In the end, the new, incomprehensible data went into a computer file, along with the Soviet sub's earlier transmission describing the strange antics of the USS *Long Beach*.

Three weeks later, Tom Wilson received an unexpected, noon-time visitor. Sitting in his cubicle work-area, with neck-high glass paneling on top of the movable wall-sections, he was engrossed in that week's PERT charts for the project he was monitoring. Things were going well enough overall, but these latest computer printouts had red-flagged three potential bottleneck sub-tasks. Any one of them could delay the predicted completion date of the project. As he began to mentally compose an inter-office warning memo to the section chiefs in charge of those sub-tasks, he became aware of the small, well-dressed man standing at the doorway.

"Yes, may I help you?"

The man remained motionless, but smiled back. "Are you Mr. Tom Wilson? I was directed here by the receptionist in the main lobby."

Tom stood and, observing that the man wore the purple badge that meant he needed no security escort, held out his hand. "Yes, I'm Wilson. What can I do for you?"

The stranger stepped forward and the two men shook hands. "Mr. Wilson, I'm Special Agent Frank Agnew. I work out of the Boston FBI office. I wonder if I could have a few minutes of your time to discuss a matter that has recently been brought to our attention." The agent reached into an inner coat pocket and flipped open a small leather wallet that displayed a shiny medal ID.

Tom was used to government visitors, and while he'd never talked to an FBI agent before, it didn't bother him to have one show up unannounced. Biosystems Lab was a government-funded outfit and he knew the FBI handled the occasional cases when classified documents got misplaced. Tom figured some poor, unlucky soul had misplaced a secret folder. It was unpleasant, but it happened. Waving to a chair, Tom said "Sure, have a seat. I need a break from these PERT charts anyway. Ordinarily I don't work through lunch, but we've got some schedule deadlines coming up and I want to be sure we don't slip."

The agent smiled again. "I won't take long, Mr. Wilson. There's nobody else around right now, so we can talk right here." He sat down and took out a pen and notepad. "Tell me, Mr. Wilson, just exactly what *is* your function here at Biosystems?"

Tom pointed at the piles of computer paper on his desk. "This is it. I'm an accountant by training, with on-the-job experience in computer data processing. I manage the weekly progress of all the little jobs—we call them sub-tasks—that together make up a whole project. Then, by knowing how they all fit together—you know, which ones require which others to be finished before they can start, and so on—I can forecast our final completion date. Or maybe I should say the *computer* can forecast the date. And even more important, it can identify those sub-tasks which are becoming critical. That means that if *they* fall behind schedule, then the *whole project* slips, too. We watch those sub-tasks *very* carefully. That is, the PERT computer program—that's 'Program Evaluation and Review Technique'—monitors things for us."[3]

The FBI agent had made some quick notes as Tom spoke, but they were about *him*, not what he did. Tom would have been surprised if he could have read them. *Wilson is intelligent, articulate, open to discussing his work, very co-operative.* He paused for a few seconds while he formed his next question and Tom mistook that for a problem.

"Say, what's this all about, anyway; have I done something wrong?"

"Certainly not, Mr. Wilson," replied Agnew, "at least nothing we know about." The agent grinned as he said that, and Tom felt better. "Look Mr. Wilson, let me be completely transparent with you. I've checked your security file, talked to your supervisor here at Biosystems, and there is absolutely no question about *you*. But there *is* a problem. As you've said, you are *not* a scientist. You've had no formal training in physics, advanced math, or electronic systems. An intelligent, well-informed person, yes, but lacking knowledge in certain specific areas of hard technology. So how do you explain the fact that your son is telling his school friends about a radiation-and-blast-activated force-shield? And saying that his *dad*, who works at the government lab, told him all about it?"

Tom Wilson sat stunned and openmouthed. Then he slowly began to turn red with embarrassment. "Oh, my gosh. Bobby must be talking about my little story! I guess I should've known a boy his age couldn't keep *that* a secret."

"'Little story,' Mr. Wilson? Tell *me* about it, will you?" said the agent.

Tom quickly filled the FBI man in on Bobby's fear, and how he had handled the situation. The agent listened quietly and made some more notes. "Well, I can understand how a father faced with a situation like that might make up such a tale. Of course, you *did* make it up, didn't you? Or *did* you hear someone here at Biosystems speak of such a force-shield?"

Tom laughed. "*Of course* I made it up. There's no such thing as a force-shield!" Then he peered closely at his visitor. "It *is* impossible, isn't it?" It was the agent's turn to laugh.

---

[3]This is a science fiction story, but PERT is real. If you are interested in just what Tom Wilson does at Biosystems, you will find a detailed discussion of PERT (including an amazingly simple computer code that implements PERT) in my book *Duelling Idiots*, Princeton 2002, pp. 71–80 and 252–257. The math of PERT is pure arithmetic, understandable by a bright grammar school kid (like Mr. Wilson's son, Bobby).

"Well, I'm no scientist, either, Mr. Wilson, but naturally, if there *was* such a thing, it would be a clear security violation to talk of it. Wouldn't it?"

Of course Mr. Wilson agreed, and the agent departed. But Mr. Wilson thought about the curious visit for a long time after that.

The three cars carrying the visiting Soviet scientific delegation sped along the nearly empty road. The scientists had been pleasantly surprised at the ease with which they had received State Department approval to attend the Montana conference on energy conservation. Two bored KGB agents, posing as delegates, sat in the car bringing up the rear as the hot, lazy desert miles pounded by.

Then the delegation encountered some detour signs, and they had to change their route. Soon after that they, and the rest of the traffic in both directions, were stopped by military police as several heavy earthmovers rumbled across the asphalt. The KGB men could see, about 200 m off to the left, down in a gully, a large group of men digging in the ground. Observing that they were being ignored by the police, the agents took some photos with their concealed minicameras. Things *were* boring, and it would show their supervisors back home that they were ever vigilant. Anyway, it wasn't smart to return from abroad with unexposed film.

Three weeks later, after blowing up the fine-grained film, the Moscow photo-interpreters were puzzled once more by what they saw. Nearly one hundred men were digging a trench precisely 30 cm wide and, from shadow measurements and the quantity of excavated dirt visible, at least 10 m deep. Fourteen hundred meters in length and straight as a laser beam, it was far too narrow for missiles. In frustration, the Lubyanka people finally consulted with their counterparts in the GRU, located in the new building for the First Chief Directorate just outside of Moscow. Military Intelligence/GRU is a division of the Soviet General Staff, and while under the thumb of the KGB, it is operated mostly by knowledgeable military men, and not the political hacks, goons, and bumblers of the KGB itself. But even the GRU was stumped by the trench photos. After much debate that led nowhere, the data went into the computer files with all the other nutty American activities. By now the Russians were getting new, crazy reports almost daily. They all went into what had become known as "the lunacy files."

A few days later, deep in the bowels of the Pentagon, a clerk/typist with an unusually high security clearance prepared a memo. To be hand-carried by an Air Force colonel to each of the 21 men and 2 women on the distribution list, it could be read only in the colonel's presence, and no note-taking would be allowed. Once all had read the document, the colonel was personally responsible for destroying it. After typing the memo, the clerk removed the typewriter ribbon[4] and placed it in a special, unmarked canvas burn-bag. Two hours later, the bag and the ribbon in it flashed into vapor as the colonel dropped them into a plasma-jet incinerator.

Classification: MOST SECRET/EYES ONLY
To: Project BUBBLE Oversight Committee
Subject: Resolution of apparent security violation
From: Office of the Joint Chiefs of Staff

(General W. K. Alison)

The initial fears that Project BUBBLE had been compromised have been shown to be incorrect. The reports of a security leak have been shown to be false. The reports of a security leak at Biosystems Laboratory were based on an occurrence which has been fully and satisfactorily explained by the report of one of our sister organization's field agents.

Indeed, the nature of this occurrence is such that it has been determined to be a higher quality misinformation operation than was the one originally planned. This fortuitous occurrence has moved the time-table up on the Project by several weeks, and this has been determined to be desirable.

All previous misinformation operations have proceeded without mishap, and additional operations are now being implemented on schedule: they are operations BLUE FLICKER, FLAME WHEEL, and KILLER BIRD. Operations SEA BEATER and BALL BLOWER are scheduled for de-deployment on schedule. Operation RAZOR TRENCH will be allowed to continue for one additional month to provide our friends with additional satellite photo opportunities.

All test activities for Project Bubble are on schedule. Four days ago the test bubble at the Nevada site successfully contained a 10-megaton explosion with zero leakage, and thus would have protected its interior against a similar weapon applied to the outer surface of the bubble. This total containment of all thermal, x-ray, gamma, and neutron radiation was achieved with a twenty-seven percent safety factor in the field-stress overload tensor. Assuming Presidential approval, the first operational bubble will be installed over Washington, DC next month. The cities of New York, Boston, Los Angeles, and St. Louis will also be provided with bubbles within the following 6 weeks. Each of these operational bubbles will be powered by two independent, enhanced statopulse inverted-field-mode generators,

[4] The use of a typewriter and its ribbon may at first seem to date this story, but I suspect typewriters with ribbons could actually still be used for documents so highly classified that you want to keep them *off* of digital devices that could potentially be hacked. The last time I was in the Pentagon was more than 50 years ago, so this is just my "best guess."

each thirty percent more powerful than the power device used in the Nevada test. This should provide a bubble radius of approximately 17 km, with an estimated blast protection of 15-megatons.

END

Twenty-five miles to the southwest of Moscow, high on a bluff overlooking the Moskva River, sits the peaceful, artificial village of Zhukovka. It is the home of the Satellite Services Department of the First Chief Directorate of Soviet Intelligence in the KGB. Major Nikolas Gorki, however, had no time to enjoy the natural beauty of the dacha's surroundings. As he carefully settled his enormous bulk of almost 300 lb into a specially reinforced chair, Major Gorki pondered his new, urgent assignment.

A rarity in the KGB, he combined a superior intellect with a ruthless nature. Many of his colleagues had one or the other of these qualities, but few had both. He was a problem solver, and he wasn't used to losing. This new puzzle would be no exception. He had complete folders on every entry currently stored in the lunacy files, and a set of photographs of unmatched technical merit for almost all of them. Besides, he had another reason for pursuing victory. While he knew from bugged embassy recordings that his American counterparts were well aware of his considerable talents, they were also quite fond of referring to him as "fatso Porki Gorki." Discovering the secret of this American game would be his revenge for such gross disrespect!

The Americans were up to *something* and hoped to mask it with a cloud of distractions. Who notices the green pear in the basket of green apples? Major Gorki smiled with quiet admiration for the shrewdness of the concept. It was a very clever form of security blanket. But he still had his basic problem.

Where was the *real* threat, hidden among all these carefully provided false targets?

Could it possibly be this long, impossibly thin, deep hole in the ground? Major Gorki carefully examined the photos of the trench with a magnifying glass, while he slowly smoked a fat Turkish cigarette. It tasted sharp, bitter, and the acrid smoke stung his eyes. It was a heavy challenge to smoke such a terrible cigarette, but it was from the challenge that came the pleasure. He pushed the trench folder aside after a while though, with a long sigh. Who could tell?—*maybe* it was a legitimate latrine, but such a hole would take the relatively few US Army troops in Montana 20 years to fill. And it was a bit too close to a public highway for *that* particular use. He took a new folder from the pile, while making a mental note to return to the hole later.

He began to laugh after about 30 s of reading. A force-field bubble, impervious to ICBM hydrogen bomb explosions? A tale overheard by some agent, no doubt drunk, while lurking in the shadows at a children's school? What utter rubbish! The Americans had slipped-up on this one—it was just *too* silly. More likely still, it wasn't even a plant, but rather the result of a child's wild imagination. Oh, it was difficult to fool a man who had dedicated his life to mastering the art of deception as an instrument of national policy!

Tossing the folder aside with a disgusted grunt, Major Gorki gave serious thought to recommending dumping all the lunacy reports. But that would be admitting defeat. The answer was in here someplace. He opened the trench folder again, and peered harder than before at the mysterious hole in the ground. Could it possibly be for the acceleration tube of an antiballistic missile, or a particle-beam weapon? *Yes*—that might make sense. Fatso Porki Gorki reached again for his magnifying glass.

He'd find the answer!

12

Computers and Society

Do you remember your reaction when you first read the 1949 novel *1984* by George Orwell?[1] I'll bet it was similar to mine: utter disbelief that such a nightmarish political system could *actually* take hold. The shocking world of *1984* would require, you would think, that a majority of people had simply lost their minds! And how likely *was that*? Well, history tells us that such things really *do* happen.

All it takes are politicians who, even if suffering from severe mental illness (what is politely called, in medical lingo, *extreme neurological impairment*), have nevertheless achieved master-level skill in the endless telling and re-telling of (in the words of Hitler's propaganda guru, Joseph Goebbels) "the big lie." In such politicians it is often the case that you can actually see, in their public behavior, that there is a significant gap between the world in which they privately exist and the world of reality. This has, in recent years, prompted outcries, *from all across the political spectrum*, for the development of "cognitive tests" to evaluate the mental acuity of the seekers of high political office.

While such concerns are serious, they are not new. Back in the 1970s I had observed such disturbing behavior in a number of politicians of those days, and wondered then what could be done to address the issue. Science fiction struck me as the perfect medium in which to propose one possibility and, while it has been nearly 50 years since "Qualification Test" appeared in the December 1979 issue of *Analog Science Fiction*, I think the approach described in the story could still be of possible use today.

[1] The pen name of Eric Blair (1903–1950), who was also the author of the equally thought-provoking 1945 novel *Animal House*.

© The Author(s), under exclusive license to Springer Nature Switzerland AG 2026

P. J. Nahin, *The Art of the "What-If?"*, https://doi.org/10.1007/978-3-032-15618-1_12

"Qualification Test"

"Damn it, *to hell*! Get over here, MacDonald. I can't get this cruddy program to run. Give me a hand, will you!"

Melvin Schluck, a large, loutish, somewhat ugly man in his late 50s, sat before the video screen of a PerComp personal home computer. He was desperately trying to get it to run a BASIC language program to add the first 200 positive integers. All that was happening was the display of the message "ERROR 53 IN LINE 30." His special aid, Professor R. Q. MacDonald, on leave from the Computer Science Department of the State University, hurried over. He was dragging down a 250-dollar-a-day consulting fee, and was eager to please. "Yes sir, Governor. Let me list your program and I'll take a look at it."

MacDonald quickly punched in the system command to scroll the six line program onto the PerComp's video terminal display. The screen filled rapidly with bright white alpha-numerics as the machine's memory unveiled its contents. He groaned to himself (the idiot *still* didn't know how to write a FOR-NEXT loop!) but with a supreme effort, he kept his thoughts silent. MacDonald paused for a few moments—he wanted the Governor to think the error was subtle enough to stump the expert, too. Then, finally, he patiently explained the problem, and after listening to the fool cackle with pleasure when the program at last ran, MacDonald excused himself to go the bathroom. He left with what he hoped wasn't too obvious a relief.

When he got to the john, MacDonald slowly loosened his tie, and then splashed cold water on his face. The only reason he was involved with the incompetent old bastard was the money. Ironically, he needed it because his salary at State U., just barely at the subsistence level, was low precisely because of the Governor's ultraconservative fiscal policies. With no broad based taxes, the major portion of state revenue came from oppressive gambling, property, and liquor taxes that barely generated enough cash to keep highway potholes filled. "Christ," he murmured to himself, "talk about being between a rock and a hard place. I don't make enough money to get smashed so I can forget I can't afford to get smashed!"

"What's the matter, Ron, can't you take working for old Mel? You know, you haven't got a chance, *not a chance*, of teaching that old dog even a single new trick!"

MacDonald looked wearily at his good-natured taunter, who was emerging from a stall while cinching up his belt. It was Jack Kraus, a professor in the electrical engineering department at the University. He and Jack were friends, but Jack had offered his training services to Reggie Smyth first. So MacDonald had signed on with the Governor (but only for the money, as he repeatedly told himself). "I tell you, Jack, I don't see how a human being can be so goddamned, bloody stupid and still be able to talk and walk at the same time. The man is a mental aberration, with spaghetti for brain convolutions. How in hell did he get elected governor *five* times?"

Kraus shrugged his shoulders and turned away from the washbasin to dry his hands with a paper towel. "Ron, the fellow may be illogical, but he's a better con man than either you or I will *ever* know. Being intelligent, under the old rules, didn't count—getting votes was the *only* name of the game. Hey, look, cheer up, will you! You're not alone, you know. My man is so emptyheaded that you could put a bullet through one ear and out the other and his IQ wouldn't perceptibly change."

MacDonald couldn't help but grin. "Reggie is really that bad? He seems like a decent sort of guy. And actually if he passes the test and gets the Democratic nomination, I was thinking he'd probably get my vote."

"Look, Ron," said a now serious Kraus, "Reggie is nice like a puppy dog is nice. And not half as smart. He does *only* what his staff of Harvard and Princeton brain-boys tell him, even though he doesn't know what in hell they're talking about seventy percent of the time—you want a person like that for senator? I rather have a clever turkey brain like Mel! But Reggie isn't going to pass, any more than Mel is."

"Yeah, the hell of it is I know you're right, Jack. But what's going to *happen*! *Who's* going to get the nomination for *either* party?"

"Beats me, Ron—I'm just here for the bucks! I figure both parties will eventually get someone to pass the test. But they won't be *our* guys! At least we'll be assured of reasonably intelligent, logically thinking candidates. But right now—I don't have the slightest idea who they might be."

"Good evening, ladies and gentlemen. This is Eric Peters, reporting *live* for the Broadcasting Services Unlimited Network at the first qualification problem solving tests to be held under the new federal election rules. With me tonight, as guest and expert commentator for BS Unlimited, is Dr. Andre Barbow. Dr. Barbow served as the special consultant to the congressional committee that drafted the original bill that eventually became the Federal Officeholder Qualification Law. As such, and as a computer expert, Dr. Barbow will interpret tonight's historic event for us. Welcome, Doctor, and thank you for being here with us tonight."

"Thank you, Eric. This *is* a historic occasion, and it is my pleasure to be able to add my impressions to the evening's festivities."

"Tell us, Dr. Barbow, what are your feelings on the possible outcome tonight? Does.

Governor Schluck have a reasonable chance of passing the test for the Republican nomination for the Senate race? Does Reginald Smyth have a chance for the Democratic nomination?"

"Well, Eric, that's hard to estimate. Both men have been, I understand, in intensive training for the test. But you have to remember that both have never had any formal education, either in college or in their later years, in the hard sciences or mathematics. Their training has been limited to the traditional courses taken by lawyers. That might hurt. But I'll tell you, Eric, they're tough men, with heart and guts. Adversity is nothing new to them. I'm sure we'll see a real show of courage tonight!"

"Right you are, Doctor, and we'll be continuing our conversation with you shortly. We'll be right back, ladies and gentlemen, after this brief but important announcement."

CUT TO A GROUP OF SKIMPILY CLAD DANCERS HIGH KICKING AROUND A TEN-FOOT-HIGH BOTTLE OF BEER, SINGING:

Pasockee, Pasockee, Pasockee Beer,
That's the drink that makes you cheer!
Light and mellow, smooth and dry,
You'll like Pasockee, so give it a try!

"Eric Peters here again, ladies and gentlemen, back with Dr. Andre Barbow. Doctor, maybe it would help give our audience a better perspective on what they'll be seeing tonight if you quickly review the history of the Federal Officeholder Qualification Law."

"Good idea, Eric! It has long been recognized, even as far back as ten years ago, that the introduction of the home microcomputer has led to a revolutionary alteration of the meaning of education in this country. Since those pioneering days of the early 1970's, when the Intel Corporation introduced the first commercial 'computer on a chip,' we have come to expect that the average person, to be called 'educated,' should know at least one high-level computer programming language, like BASIC, FORTRAN, or PASCAL. *And*, and here's the rub, Eric, he or she should know *how to solve problems* with the aid of a computer. The development of the Federal Officeholder Qualification Law is the logical conclusion of that realization—that is, to be a candidate for higher elective office, a person must openly demonstrate that he possesses such intellectual, decision-making, problem solving skills."

"It *is* incredible, wouldn't you agree, Dr. Barbow, how the home computer revolution has seemed to happen almost overnight?"

"Well, actually that gets a yes *and* a no, Eric. The growth process has always been there, but its pattern has been geometric, a *very* difficult one to perceive in its initial stages. I like to compare this kind of situation to a 'lilly pad in a pond.' Imagine you have a very large pond, with a very small lily pad in the middle. Each hour the pad doubles in area. For a long time, it's not very big. But, eventually it gets to the point where it covers one-fourth of the pond's surface. One hour later, the pond is half covered. The very next hour, the entire pond disappears! The pond vanishes, at the end, with almost awesome speed. And this is exactly the process by which the home computer market has grown."

"It is this phenomenal growth that has led to the Federal Officeholder Qualification Law, then?"

"Absolutely, Eric. Because today, the average person is algorithmically literate. Let me put all this another way—the impact of the personal home computer is comparable to that of the hand gun in the Old West. The gun equalized man's *physical* differences. The private computer has gone a long way to doing the same for his intellectual differences. So much so, in fact, that it's now reasonable to expect, indeed insist, that our public servants *all* be the best that can be had."

"No question of that, Doctor, that's the American way! But right now we are going to take a break for another informative announcement from our sponsor."

CUT TO GROUP OF SKIMPILY CLAD DANCERS—

Pasockee, Pasockee, Pasockee Beer,
The beer of the future is already here!
Goes great with nuts, pretzels or cheese,
Drink Pasockee Beer, it's sure to please!

"Thank you, Dr. Barbow, for staying with us. But let's go back to the issue we touched on, just briefly, a few minutes ago. Not everybody has 'grown up' with the computer revolution. What about the experienced politicians, those men and women who have been around so long their early education missed the home computer revolution. Aren't they being discriminated against?"

"Yes, Eric, that *is* a serious problem, and we recognized it early in the game, when preparing the initial drafts of the Law. And that's why *all* people, present officeholders or not, who 'try-out,' will be given 'bonus' points on the basis of their age. The older you are, the more bonus points you get."

"But Doctor, won't that encourage *really* old people to run? I mean, they'll get a lot of bonus points. Isn't that right?"

"Yes, Eric, it is, but there's a trade-off involved. You see, a person who is very old will probably do so poorly on the actual tests, themselves, that the bonus points will not be sufficient to score at a qualifying level. But anybody with reasonable intelligence, from twenty-five, to seventy-five years of age or so, should have no real trouble."

"Thank you, Dr. Andre Barbow, for those penetrating insights. But we must break away from this interesting line of discussion, as I see the candidates are coming on stage now, and the tests are about to begin! So here we go, ladies and gentlemen, the next three hours will be historic. Pick your favorite, place those little side-bets you might have, cross your fingers and—here we go!"

"This is Eric Peters, ladies and gentlemen, wrapping it up for you after tonight's historic events. Who would have guessed it—a dead heat tie!—with both Governor Schluck and Reginald Smyth getting thirty-seven points, each, *including* twenty 'bonus' age points. Thus, neither man qualifies for the candidacy of his party, which would have required a minimum of sixty-nine points."

"But, ladies and gentlemen, I want you to take a close look at the two disappointed men out there on center stage. Disappointed, yes. But defeated in spirit? No! What a marvelous lesson in personal integrity and good sportsmanship the Governor and his colleague are giving our young viewers. The Governor has his arm around Reginald Smyth's shoulders, and appears to be trying to buck up the spirits of his fallen comrade. I just wish we could get a mike down there so we could all hear what I'm sure are inspirational words. Words that would warm the hearts of all Americans!"

"Goddamn it, Smyth, we've just been royally screwed! And did you see that pompous bastard who did it—Barbow—up there in the press box area? Boy, would I like to fry his butt!"

"Well, Governor, forget it. It's over for both of us. So, forget it."

"You're a sniveling, spineless doormat—you know that, don't you Smyth? The hell I'll forget! Look, when I came in here tonight I saw where that jerk Barbow parked his car—are you with me or not?"

"What do you mean—'with you'? What are you going to do?"

"Smyth, all you've got to do is stand look-out for me, understand? Now, let's get out to the parking lot before we get mobbed. I wonder how far Barbow will get tonight, after I let the air out of all four tires! And put a little sugar in the gas tank, too!" Schluck rolled three sugar cubes in his hands that he had swiped from a supermarket a few days earlier.

Reggie was shocked. "Would you *really* do that Governor? That could wreck his car!"

Schluck cackled with pleasure. "Would I do it? Follow me, Smyth, and watch!"

"Good evening, ladies and gentlemen. This is Eric Peters, your host on *Celebrity Interview*. This week's guest on the BS Unlimited Network is Dr. Andre Barbow, founder and president of the controversial Famous Programmers School. Dr. Barbow's professional credits are many. One, that you may recall from last month's election night coverage, was that of special consultant to the original congressional committee that put together the basic text of the now famous Federal Officeholder Qualification Law."

"We will question Dr. Barbow about his latest venture, right after this important commercial message."

CUT TO FOUR PERKY-LOOKING HOUSEWIVES TOSSING ROLLS OF *SPONGY* PAPER TOWELS INTO THE AIR AS THEY SING:

"Forget those spills, forget those drips,
No more fear of kitchen slips!
The scientific towel is the one to try,
Throw it on a puddle and it'll be dry, dry, dry!"

"Okay, folks, I do hope you'll seriously consider our sponsor's timely message. Now, let's talk to our guest, Dr. Andre Barbow. Welcome to *Celebrity Interview*, Doctor."

"Thanks, Eric, it's certainly nice to see you again."

"Yes, Doctor, the last time we met was on the occasion of the very first use of the Federal Officeholder Qualification Law. Please tell our audience, Doctor, of how you now evaluate that night, and all of the other qualification results across the country, since the Law became effective."

"Well, Eric, things *have*, of course, been in a bit of turmoil. As I recall, only five of the House candidates qualified, none of them incumbents, and no Senate candidates qualified at all. Certainly a rather disappointing outcome for the state party structures! Of course, we must also look on the brighter side of it—Capitol Hill will receive a fresh infusion of new blood. And, I would like to think, new and better brains!"

"Yes, Doctor, it will *if* computer-qualified candidates can be found. What happens to Congress if the two parties *can't* find such candidates?"

"I understand your concern, Eric, but it really isn't a problem. First of all, there is the 'grandfather' clause in the Law. That will assure Congress will not collapse from a lack of bodies to fill the seats."

"The 'grandfather' clause, Doctor? Just what is that?"

"The 'grandfather' clause, Eric, allows for the continuation in service of any congressperson, past the normal term of office, *if* no new candidate for election can be found. This is, admittedly, probably subject to challenge in the courts as unconstitutional. But I feel that by the time it reaches the Supreme Court, the danger period of crisis would have passed. So it's just our first level safety valve."

"Because, you see, even without the 'grandfather' clause, Eric, we would *still* have no problem. After all, there are plenty of bright, young unemployed Ph.Ds in America who can program-up a storm. I believe there's a bountiful pool of qualified people—we just have to get them to turn their interests to politics! And, of course, the pay isn't bad, either!"

"Yes, I suppose you're right. But those people would all be neophytes at politics, with absolutely no experience in how things are done in Washington. Can the country afford to lose that vast backlog of knowledge?"

"No, Eric, it can't! The 'grandfather' clause, and the unemployed Ph.D resource, are really just stopgap measures. Our real hope still lies with the 'old pros.' We just have to 're-tool' them, so to speak, to bring them up to speed computer-wise, and into the technological age!"

"Ah, Doctor, I'm glad you mentioned that! That brings us to your recent founding of the Famous Programmers School. Just what is that, anyway? But before you answer that question, Doctor, let's break for a message from one of our sponsors."

CUT TO PERKY HOUSEWIVES—

"Milk and water and juice and punch,
SPONGY towels drink them for lunch!
SPONGY towels, the scientific one,
A pound of SPONGY soaks up a ton!"

"Okay, Doctor, let me repeat the question. What is Famous Programmers School, your newest professional activity?"

"Eric, I'm so proud of Famous Programmers I could just bust the buttons off my vest! We have assembled some of the top minds in computer science around the country who then provide correspondence training to potential officeholders."

"Reginald Smyth—you may remember him from last month—is one of our charter membership students. And doing quite well, too, I understand! With the aid of our instructors, he has, in just *three* lessons, learned how to avoid infinite loops and, now get this, Eric, this is really just impressive as all heck, he has *mastered* the technique of inserting comment statements into his program listings to document the logic of his computer code. This is, as you may well know, contrary to all previously accepted political behavior, which says don't put *anything* in writing. Who says you can't teach an old dog new tricks!"

"Yes, that all sounds very good, Doctor, but surely you are aware of the criticism that has been directed at you. Namely that, first, you were a principal in the creation of the Law, and second, that you are now personally gaining from its enactment by establishing Famous Programmers. In effect, you've created a captive audience!"

"Eric, that's just pure, vicious, slanderous poppycock! I have merely pursued the American way of capitalism—I've identified a need, and I'm meeting that need. My previous involvement with the new Law is totally irrelevant."

"Certainly no one is accusing you of illegal actions, Doctor. Not yet, at least. But there *is* a grass-roots movement, headed by Governor Schluck, demanding an investigation of you and your Famous Programmers School. How do you answer those critics?"

"Eric, Governor Schluck is on a personal vendetta against me. First, he didn't qualify as a Senate candidate. And then there is that disgraceful episode after the tests where he was caught—red-handed!—vandalizing my car! If he didn't have the State Police in his back pocket, he'd be doing thirty days in the County Jail for criminal mischief! It's all sour grapes on his part, and I pay him no attention. And nobody else should either."

"Yes, Doctor, that *was* a curious incident with Governor Schluck. But I see we have time for just one more question. You say Famous Programmers is a *correspondence* school. How can you do that—how do the students get any 'hands-on' experience, so to speak? Isn't teaching computer operations, by *mail*, as impossible as it would be to teach how to be an *airplane pilot*, or a *brain surgeon*, by correspondence?"

"Normally, Eric, using all previously accepted concepts of education, your skepticism would be warranted. We, however, have developed, with the assistance of some of the top educational experts in the country, a broad spectrum of techniques to avoid the very problem you mention. We have put together a comprehensive learning package unique in its total impact. Yes, a truly *integrated pedagogical system* that presents an aggressive, hard-hitting, dynamic sequence of coordinated, interrelated lesson plans!"

"But Doctor, I don't think that actually addresses my question. Just what, *specifically*, do—what? No more time? Well, we're out of time, Doctor. Possibly we can pursue this on another day. Many thanks to our guest, Dr. Andre Barbow, President of Famous Programmers School. This is Eric Peters saying so long from *Celebrity Interview*, another in one of the BS Unlimited Network's educational presentations."

> CUT TO WHITE-HAIRED, CONSERVATIVELY DRESSED, VERY SOPHISTICATED LOOKING MAN IN MOD GLASSES. HE IS SEATED IN FRONT OF AN IMPRESSIVE COMPUTER, WITH MAGNETIC TAPE DRIVES SPINNING, LINE PRINTERS OUT-PUTTING PAPER AT AN ENORMOUS RATE, AND A TESLA COIL ARCING NEARBY. HIS RIGHT HAND RESTS ON A TELEPHONE RECEIVER. AN ATTRACTIVE, OLDER COUPLE, EACH HOLDING A GLASS OF WINE WHILE READING A PROGRAM LISTING, IS IN THE BACKGROUND. THE SEATED MAN SPEAKS IN A LOW-KEY VOICE, AS SOFT MUSIC PLAYS:

"This message is for the very special man or woman. You are successful, a person who has met and mastered the challenges of life. You are established. You are the result of hard work, steadfast perseverance, and dedicated effort. And yet, and yet—there is one challenge left to you."

"Congress."

"Yes, precisely because of your high level of seniority, you missed the computer revolution. Due to the new Federal Officeholder Qualification Law, you cannot qualify. But this is merely a temporary difficulty."

"The Famous Programmers School, founded by one of the men who wrote the law, can help you meet this final challenge. And with it, the prestige and honor of going to Washington can be yours, too. A Famous Programmers School diploma, earned in the privacy of your own home, is the mark of distinction, the symbol of a person who is a doer. Not a quitter, and certainly not a loser."

"Washington, with its power and its night life, is as close as your telephone (MAN GENTLY PATS TELEPHONE). *Call this toll-free number now* (NUMBER FLASHES ON SCREEN) *and arrange for a no commitment, no charge, evaluation test. No one will call."*

"Sign-up for a Famous Programmers session, however, and the future will *call. Don't delay, call today."*

"A Congressional seat, with your name on it, could be just around the corner!"

A color television set explodes in a burst of sparks as a paper weight smashes through the screen. "That damn SOB! I'll see Barbow rotting in jail before this is over!"

A pale, shaking Mel Schluck stood screaming at his now smoking TV. His face was contorted in a rage of frustration, and his fists clinched and relaxed reflexively. His blood pressure soared to 290 over 180, and the unsuspected congenital aortic aneurysm ballooned dangerously.

Schluck moved around his desk to reach for the telephone. “By God, this is just too much! I’ll call a press conference and stick it to that bastard *in print*!” The arterial wall was stretched paper-thin now, and a trickle of blood started to seep through the yielding membrane. He began punching in the number of the local newspaper. He got the first three digits in, and then as his mind thought the fourth, the artery went forever past its Hooke’s limit. In a chest wrenching, gut shattering knife of blinding pain, the wall burst open and the blood of Melvin Schluck splattered off his lungs, into his abdominal cavity. Dead before he hit the floor, the Governor had failed his last test.

The coffin was finally alone, waiting for its last, short journey of 6 feet straight down. A cold, steady wind blew over the open yaw of the grave-pit, scattering petals from the flower wreaths draped over the box containing Schluck’s remains. The crowd of mourners was already spreading out over the immaculate cemetery lawn, heading for the parking lot. Eric Peters, covering the funeral for BS Unlimited Network News, hurried after the familiar figure a few dozen feet ahead.

“Dr. Barbow, please wait!”

The rapidly striding figure turned, stopped, and waited for the approaching reporter. “Hello, Eric, *thought* I recognized you back at the services. How are you?”

“Fine, Doctor, thank you. But I’m surprised to see *you* here. After all, you and Schluck were hardly friends. In fact, except for his sudden death, he probably was going to cause you quite a lot of trouble. Why *did* you come, Doctor?”

“Eric, Governor Schluck’s sudden passing *proves* that he was under enormous pressure. His strange behavior was evidently the direct result of the strain his deteriorating health put him under. You may not believe this, but I *always* admired the man. And it was only fitting that I pay my last respects.” The reporter, rapidly taking notes, lowered his head as he wrote to hide a cynical smile. With Schluck out of the picture now, Barbow’s problems were over. Barbow stuffed his hands into his coat pocket against the cold, and continued.

“Yes, Governor Schluck, in his own unique way, was a great man. I hold no ill feelings toward him—indeed, I think his memory should be honored. This is on the record, Eric. Famous Programmers School will begin to award a certain number of full tuition memorial scholarships in the Governor’s name. Maybe two or three a year. We’ll call the recipients ‘Schluck Scholars’!”

“Well, Doctor,” replied Peters, “that certainly is magnanimous of you. Also very politic. Any residual thought of pushing for an investigation of Famous Programmers School should be defused by *that* move.”

"Come, come, Eric, there's no covert motive involved. Merely a gesture of honest emotion to remember a fine man. Just think—someday most members of the House and Senate could be graduates of Famous Programmers!" Barbow's face glowed with pleasure as he dwelled on the thought. The two men continued their walk away from the grave, into which the grounds staff were now shoveling dirt.

"And think of this, Eric, think of *this*. We may even live to see the day when the President of the United States will be a Schluck Scholar! Now, wouldn't *that* be *something*!"

Eric Peters turned, almost involuntarily, to take one, last look at Schluck's grave. A slight shudder went through his body. *Jesus,* he thought, *the President of the United States a Schluck Scholar! You may be dead and buried, Governor, but I don't think you'll be forgotten. Nosir, not forgotten by a long shot!*

After finishing "Qualification Test" you might think you have an idea of who the real politician in the current news was the inspiration for Mel Schluck. But you *have* to be wrong because, you'll recall, the story was written a half-century ago when your Schluck was not yet even in politics. If you continue to insist that your guy is an absolute dead-ringer for the story's Schluck, all I can say is that the world will always (alas) have plenty of Schlucks and that is why *your* Schluck isn't *my* Schluck (who, in fact, is a composite of multiple Schlucks who were prominent in the 1970s).

13

Far-Out Physics

When most people hear mention of Einstein's famous equation, $E = mc^2$, they immediately think of *nuclear weapons*. That is, either of a *fission* uranium or plutonium A-bomb, or of a vastly more powerful hydrogen *fusion* H-bomb (an A-bomb is required, just to *ignite* an H-bomb, to start it—and everything else for miles around—on the way to oblivion), both of which convert mass into energy. The energy (E) associated with even a very small mass (m) is *huge*—*huge* because the speed of light (c) is very large to begin with, and then it gets *squared*! The resulting huge E appears as an awesomely enormous fireball explosion that, for a period of time, replicates on the surface of the Earth physical conditions that exist in the interior of a star.

What doesn't commonly occur to most people is the idea of running the Einstein equation in reverse. That is, instead of putting m into the equation and calculating the associated E, let's put E into the equation and calculate the associated m. Running $E = mc^2$ backward explains the mathematics behind a common claim you can find in just about any book on the Sun: *each second, four million tons of the Sun's mass is totally converted to energy*. That usually grabs the attention of readers, but where does that impressive number (4,000,000 *tons every second*) come from?[1] Let me take you through the calculations—it's all just arithmetic—because then, when you read this chapter's story ("What *Really* Caused the Energy Crisis," which appeared in the July 1978 issue of *Analog Science Fiction*), you'll understand where the numbers in that story come from.

---

[1] The Sun has been doing this mass-to-energy conversion for billions of years, and will continue to do that for billions more years. The Sun, although just an average-sized star, is nevertheless still pretty *big*! (It has a diameter of 865,000 miles.)

© The Author(s), under exclusive license to Springer Nature Switzerland AG 2026

P. J. Nahin, *The Art of the "What-If?"*, https://doi.org/10.1007/978-3-032-15618-1_13

Imagine the Sun is at the center of a spherical surface with radius 93,000,000 miles (the radius of Earth's orbit). The Sun is continually radiating energy into space in all directions, and it is not difficult to measure the power (energy per unit time) incident on Earth's surface (using, for example, solar cells). The result is about 1200 watts per square meter, at the equator.[2] Since one mile is 1609 m, the radius of Earth's orbit is

$$\left(93{,}000{,}000)(1609\right) = 1.5 \times 10^{11} \text{ m},$$

and so the total energy output of the Sun, each second, is given by (remember, the surface area of a sphere with radius r is $4\pi r^2$).

$$\left(1200\right)\left(4\pi\right)\left(1.5 \times 10^{11}\right)^2 = 33{,}929 \times 10^{22} = 3.4 \times 10^{26} \text{ J/s}.$$

The speed of light is $c = 3 \times 10^8$ m/s, and so the mass of the Sun being converted to energy each second is

$$m = \frac{E}{c^2} = \frac{3.4 \times 10^{26}}{9 \times 10^{16}} = 0.38 \times 10^{10} \text{ kg} = 3.8 \times 10^9 \text{ kg}$$

A metric ton is 1000 kg,[3] and so we have

$$m = \frac{3.8 \times 10^9}{10^3} = 3.8 \times 10^6 \text{metric tons}$$

for the mass of the Sun converted to energy, each second, which is pretty close to the commonly stated four million tons per second.

The story for this chapter was motivated by events that came (in the 1970s) to be called the "energy crisis," events whose origin can be traced back at least as far as to the previous decade. At that time oil imports to America were strictly regulated, to insulate American oil producers from foreign competition. That resulted in the depletion of America's own oil reserves at a rate faster than would otherwise have been the case. (You would think that would have been an obvious result, wouldn't you?) So, America's oil reserves

[2] All the calculations we will do here will be in MKS units (meters, kilograms, seconds, joules, watts). The joule is the MKS unit of energy, and one joule per second is one watt. So, energy from the Sun is arriving on Earth at the rate of 1200 joules per second per square meter.

[3] There are 2.2 pounds in a kilogram, so a *metric ton* is 2200 pounds, 10% larger than the *English ton* of 2000 pounds.

dwindled, and then the chickens came home to roost in late 1973, when the Arab members of the Organization of Petroleum Exporting Countries (OPEC) cut oil production as a way to protest the support provided by America to Israel in the Yom Kippur War.

In response to OPEC's decision, President Richard Nixon put in place a program of gasoline rationing, to both make up for the decline of foreign oil imports to America, and to conserve America's oil reserves. The resulting nightly images on television screens, showing long lines of cars at gas station pumps, stunned people, and thus was born the "energy crisis."

As a way to respond, long term, to the then obvious danger inherent in America's dependence on oil, foreign *and* domestic, environmental activists started championing the development of alternative energy sources, like solar, wind, and tidal power. One energy source I *don't* recall being endorsed, though, at least at first, was that of nuclear power. That was because nuclear energy brought a lot of baggage to the table, such as the problem of radioactive waste disposal and the fear of a nuclear reactor core meltdown.[4] But what-if, I wondered then, the "crisis" became so severe that atomic energy started to look acceptable? What-if people started *demanding* nuclear power plants? Pursuing that thought, I took it one step further: what-if there were another reason for the need to develop nuclear power, a reason that would get people who had power and money on-board along with the environmental activists? What-if, for such people, the "energy crisis" was actually a cover for a far-darker reason?

With the math calculations that opened this chapter, and this little bit of political history, you are all set for "What *Really* Caused the Energy Crisis," a story describing a freshman college lecture that I would be willing to bet every physics professor in America would love to be able to give.

"What *Really* Caused the Energy Crisis"

Walking into his 8 o'clock freshman physics class, Senior Full Professor Doctor John Pixley had the feeling the students looked familiar. But, of course, they weren't. It was the first class meeting of the fall semester, and all the students were new to the campus. Looking around at the hushed, awed class—Pixley knew *that* wouldn't last past the third meeting—he decided it

[4]This fear inspired the 1979 movie *The China Syndrome*, starring Jane Fonda and Jack Lemon. That was the year of the Three-Mile Island (Pennsylvania) partial-core meltdown, which in fact was *far less* horrific than the later (1986) Chernobyl *total* core meltdown disaster in Ukraine.

must be the firm bodies, tanned skin, and youthful excitement that was the same each year.

If the students were new, Pixley certainly wasn't. Sixty years old—*but still trim in figure,* he thought—he was a full prof more by reason of 35 years of dedicated teaching than by publication of scholarly papers. No, Pixley had long since resigned himself to be the sort of teacher who gains influence and some sort of immortality through his students. But Pixley knew his worth. He didn't take second place in the department pecking order to any of the hot-shot publication artists, no sir.

"Good morning. This is Physics-I, just in case any of you should be in Bio-I, which meets in Room 241-B." Pixley did this routine introduction each year as his class was invariably assigned Room 241-A and a few students always got confused. Sure enough, an embarrassed young lady got up and, followed by some giggling, left. As he watched, he thought of a possible experimental proposal that might be worth submitting to NSF, assuming he could count on somebody at the National Science Foundation having a sense of humor while reading his pitch to fund a study of "The Resonance Modes in the Motion Dynamics of Two Balanced, Juxtaposed, Periodically Disturbed Hemispheres in a Uniform Gravity Field." *Fat chance of that!* he thought, as he recalled the stuffed-shirt rejection his last proposal to NSF had gotten, before the boys at DARPA—after hearing about his proposal—had stepped in and changed everything.

His attention was interrupted by the appearance of four new, red-faced men, obviously freshly emerged from Bio-I. The class got a new excuse to laugh at the expense of the confused, and Pixley noted with disappointment there was no replacement for the young lady. "Okay, let's get started. We are going to cover a lot of ground this semester, and I'll expect hard work out of you. This class will be demanding. But you will find, even if you are here just to satisfy the University science requirement, that it can be fun and you'll learn a lot about how the world we live in works. Even if you go into a non-technical career, a basic knowledge of physics will help differentiate you from the uneducated, the uncultured, and, if I may say so, the uninteresting." Pixley was only momentarily hesitant with these last words, as while he wasn't sure they were true, he was convinced they *should* be.

"You may not have thought much about it before, but you already know quite a bit about this subject. Just to survive to the ripe old age of seventeen or eighteen, you've had to learn how a lot of things in nature work. Like if you jump off a porch, you'll fall rather than float, So you know something about gravity. You've learned to get out of the way of a bicycle, so you know something about kinetic energy, and energy coupling between bodies undergoing collision."

Pixley stopped pacing back and forth as he talked, and looked at the class. They looked back expectantly. No one showed the slightest intention of asking a question. "In fact," he went on, "you even know a lot of theoretical things, too, that go beyond mere experimental observations of everyday life. For example, is there a limit to how fast things can move?" Most of the students looked perplexed at that, but several *did* put up their hands. Pixley pointed at a red-headed young man in the back of the room, and the boy responded.

"Well, I guess you must mean the speed of light. I mean, nothing can go faster than light, right?"

Pixley nodded, happy to have been given a good line to follow. "Yes, that's pretty much true. It's not *exactly* right, because sometimes things *can* go faster than light *if* they're both moving in something other than empty space, like water. But you're essentially right—nothing can go faster than light does *in a vacuum*. Now, let's try something else." Reaching into his briefcase on the floor next to the lectern, Pixley extracted a folded, dogeared color print and held it up to the class.

"This is an aerial picture I picked up years ago when I did some consulting work for the Navy. It's a shot of the old carrier *Enterprise*, with her crew—well over 5,000!—on deck in their dress whites. Anyone tell me what it means?"

The red-haired boy spoke up again, drawing some resentful stares from his classmates. *That boy is going to learn what happens to students who speak up*, thought Pixley. That old school-boy rule of "never show-off your knowledge" was in operation, and Pixley made a mental note to learn his name. He looked promising.

"The crew is all lined-up to spell out $E = mc^2$, so I'd guess that's an old nuclear-powered ship. Because isn't that Einstein's mass-energy law?"

"Right again," said Pixley, pleased. "Next to Maxwell's equations for the electromagnet field that make radio possible, Einstein's equation is probably the most impactful creation of the human mind. And you already know of it, just as part of growing up in our modern atomic powered society."

A young lady next to the red-haired boy spoke up at that. "But shouldn't you add, professor, some comments about the terrible social problems atomic energy has caused? The spread of reactors, with their radioactive waste, threatens everybody. And all just because it makes money for a special few at the risk of ordinary people!" Pixley felt a cold wave of apprehension as he recognized the tone of a young idealist speaking with all the fervor that being naïve can muster. He responded carefully.

"Of course there *are* great difficulties of the kind you mention, but so far there haven't been any major problems, except for the partial core meltdown at Tulsa a few years ago.[5] And with the Livermore Lab fusion process—have all of you heard of that?—good—soon to be coupled into the national power grid, we will then only have the tritium disposal problem."

The young lady looked unconvinced, as did many of her classmates.

Pixley knew from sorry experience that he could lose the class, permanently, right here and now, if he didn't regain control. To be challenged in the first 10 min of the first class meeting of the year, and not to persuasively counter, would be fatal. He gained a few seconds to gather his thoughts by lighting his pipe.

"Look, I know what are the prevailing ideas about the spread of nuclear power in the US. That a bunch of capitalistic, money-grubbing, public-be-damned power utilities, and rich corporate investors, found it an attractive way to make huge profits in response to the world energy crisis that first started to be publicly debated in the early '70s. Well, that is wrong! There was *some* of that, of course, but alone it wouldn't have been sufficient to result in the massive energy conservation, new energy source development programs of the last three decades or so. Something else has been at work. And that 'something' was national security! Now I know most of you are aware of how that excuse was terribly abused in the infamous Nixon-Watergate scandal, and in the even more shameful Slaver Inquiry more recently, but in this case it's true."

The class still looked skeptical, but there was interest, too, on the faces looking back at him and Pixley knew, for the moment, that he had them back in his hands. Now, to keep them there. Pixley picked up his lecture notes and dramatically tossed them back into his briefcase.

"To hell with today's lecture. I'm going to instead tell you a story that you may not believe, a story that the powers-that-be in the White House, Pentagon, and State Department will deny, but is nevertheless true. It isn't classified because to classify it would be to admit it's true, and there is always someone who would leak it, anyway. Like me. Today won't be a loss in any case—you'll learn some physics from what I am about to tell you. How many of you have heard of the government agency called DARPA?"

No response.

"Just as I thought," said Pixley, "and that isn't strange. DARPA tries very hard to keep a very low profile. DARPA, which stands for 'Defense Advanced Research Projects Agency,' is part of the Directorate of Defense Research and

[5] Remember, the core meltdown at Three-Mile Island (see note 4) was still a year in the future when this story was written in 1978. I had to *make up* a core meltdown at Tulsa for story effect.

Engineering, which in turn is part of the Department of Defense. DARPA was chartered decades ago with the basic mission of providing initial financial support for 'blue sky,' even almost crazy ideas for basic research that just might lead to new weapons technology. The idea has been to avoid an occurrence of what military people call 'technological surprise' on the United States. That refers to the possibility a potential enemy of the US suddenly comes up with a revolutionary new weapon that lets them get the upper-hand in the military stalemate we presently have."

Pixley paused for a moment, to let all that sink in—he took a few seconds to enjoy a puff or two on his pipe—and then said "And please don't think this is just paranoia, because such catastrophic weapons developments *have* happened in the past. Can any of you think of some examples?" The red-haired boy in the back responded again, but Pixley could see that now there was more interest by the class in what he had to say.

"How about the atomic bomb, professor? When the United States dropped two of them on Japan in World War 2, that must have caught them by surprise!"

"Yes, indeed, it certainly did," replied Pixley. "To the Japanese it was pure magic, although that may not be exactly the right word. They had absolutely no idea what had happened. Curiously, German documents, and interrogations of captured scientists, indicated that the Third Reich had been surprised, too. While the Germans *did* know *what* had happened, it was soon clear that they didn't know *how* it had happened. They had never been able to develop an atomic bomb program that was even remotely like our Manhattan Project. But that is pretty recent stuff. Technological surprises have actually occurred all through history, to the sorrow of the surprised. For example, you might look into what happened to the flower of fourteenth century French Knighthood at Crecy, when the knights encountered the armor penetrating arrows of the English longbow for the first time. Or even earlier, look into the tenth century battle for Constantinople, and how Russia's Prince Igor had his fleet of 10,000 ships burned to their waterlines when the Byzantines zapped them with their version of 'Greek Fire.' And, of course, the classic case of technological surprise *has* to have occurred during the Biblical battle between David and Goliath."

Pixley smiled as he watched the students furiously take notes, and then continued with "DARPA itself came into being because of a major surprise on the US—the satellite launching of Sputnik by the Russians in 1957. But I am drifting away from what I really started out to tell you. In the early '70s, DARPA was suddenly confronted with a *colossal* technological surprise from Russia. Yes, that's right, something that threatened to leave us wide open to them. The Russians were on the verge of developing a weapon system that would completely negate our ICBMs. *The perfect antiballistic missile system!* It was based on directed energy beams, consisting of space-based high-energy lasers or subatomic particle accelerators. The Russians had made fantastic strides in such devices, and just because such weapons violated the then existing ABM and Outer Space Treaties didn't make any difference. Once they had such weapons actually deployed, there wouldn't be a damn thing we could do about it."

"But professor, couldn't we just shoot their weapons down?"

Pixley fixed a fatherly smile on the student who had spoken. "With what? Their new weapon was *designed* to shoot down *our* missiles! And that's all we had to shoot at their satellite-based energy beam projectors. No, once in orbit, they would have really had us by the neck." The class was hanging on the edges of their seats now. Pixley relaxed a bit, and began to enjoy himself as he continued his tale.

"You can't believe the uproar that rolled through the corridors of the Pentagon. The careers of a lot of Generals, Admirals, and senior civilian intelligence officers came to an abrupt end once Congress got hold of this story. As it turned out, some isolated Air Force intelligence people *had* been warning the CIA about the Russian advancements, but they had been dismissed as alarmists. It was the old 'not invented here' syndrome. If the US couldn't build such weapons, why then how could mostly rural, peasant-like Mother Russia do it? Well, by God, *they did it*!"

A bespectacled, bright-looking young man in the front row, quiet up to now, raised his hand. Pixley gave a quick nod to him.

"Professor, this all reminds me of that old cowboy and Indians joke, where the ancient Indian fighter is telling how, years before, he had been surrounded by attacking Indians and how he was down to his last bullet. When asked how it turned out, he replied "I got killed!"" The class roared, and Pixley glowed. A better entrance line he couldn't have written himself.

"Quite true, young man, quite true. Here we all sit, alive and healthy, America is still here and Russia is still there. And how do I explain *that*? Could it be true that once Russia defeated us without firing a shot they just said "Okay, we won but don't worry, everything stays the same. We just wanted to show you guys we could do it."" A few snorts of laughter rippled through the class, and Pixley himself had to suppress a smile. Pixley kept quiet for half a minute, just smoking his pipe[6] and letting the class think about it. Thirty-five

[6] If anything dates this story, it's Pixley's pipe. When I was teaching, decades ago, you *might* be able to get away with smoking in a classroom, but of course today a professor smoking in a classroom would be in a

years in the lecture room had taught him that to be a good teacher wasn't just about knowing your subject. You had to put on a show, too.

"Just because the CIA blew it doesn't mean we were totally defenseless. The boys over at DARPA hadn't been sitting on their hands all those years. It seems they had been funding the research of a little-known professor in a small New England university who had developed a wild idea in the late '60s. Remember Einstein's '$E = mc^2$'? The way we used that in the second atomic bomb, the one dropped on Hiroshima, was to convert a small portion of a uranium mass into a heck of a lot of energy. The first and third atomic bombs, at Alamogordo in New Mexico and at Nagasaki, did the same using plutonium. And, of course, that's what we do today in fission and fusion reactors. Except in fusion reactors it's *hydrogen* mass that gets converted. What this professor wanted to do was *run Einstein's equation backwards!* Do you see what that means?" Pixley could see they didn't see, and this time he didn't wait for any dramatic silence to help him out.

"What the professor wanted to do was take electrical energy and literally compress it back into mass. The kicker in that was that his theoretical analyses predicted that the end-mass didn't have to have any relation to the mass that had been converted to produce the energy he started with! In fact, energy produced from ordinary chemical reactions, like burning wood, was just fine, although of course not very efficient. A wood burner doesn't put out much energy compared to a nuke plant! Anyway, he claimed he could create any of the known elements, along with new, super-heavy ones that his theory predicted; by so extending the periodic table, he claimed he could produce elements *not known to science.*"

Pixley puffed on his pipe to get it going again, and blew a smoke ring. He continued.

"One of those super-heavy elements had the theoretically calculated property of making an ultra-hard, super-heat-resistant alloy if combined just the right way with titanium. That alloy was, appropriately enough, called *Impervium.* When used to make the outer skins on our missiles, those missiles were highly resistant to the Russian ABM energy beam projectors. So, you see, we had what is called a 'Mexican standoff.' The Russians could build an ABM system we didn't know how to build, but we could build missiles *they* didn't know how to build. Counter-surprise!"

"So, both sides were in the same spot. We couldn't shoot down all their missiles, and they couldn't shoot down all of ours. For the next ten years both sides worked like hell to catch up with the other side's technology. In the end,

huge barrel of trouble.

both sides ended up with duplicate capabilities. As of right now, both sides have refrained from deploying ABM energy beams because both sides have *Impervium*-skinned missiles. But, of course, this happy ending, happy so far anyway, occurred only because of DARPA's excellent judgement in deciding to support the good professor's nutty idea. I might add that he was quite handsomely rewarded, so don't worry about him, even though he didn't get the Nobel prize—at least not yet. He *did* get a heck of a lot of money, though, along with the Presidential Science Medal. All very private and hush-hush, of course."

"Professor," came a query from the red-haired boy Pixley had noticed earlier, "that's a very interesting story, but just what does it have to do with the energy crisis? I mean that's what you started to tell us about, right? And why don't people talk about this today? It's all over, right? And why does the professor in your story have to keep quiet about his invention even after the Russians have developed the *Impervium* technology, too? What would be the point to being secretive *now*, today?"

"Those, young man, *are* the crucial questions. And the answers are really very simple. To answer the first one, just calculate how much energy it takes to make a kilogram of *Impervium* additive. In fact, I'll save you the trouble. Plug the value for the speed of light into Einstein's equation and it's not hard to calculate that it takes twenty-five *billion* kilowatt-hours of electrical energy to make a one-kilogram mass of *anything*! That, my young friends, is just one hell of a lot of energy."[7] Pixley could see that the students were struggling with the implications of that enormous number, and decided to illustrate things in a different way.

"One of the new fusion plants will run, full out, at a power level of 10,000 megawatts. You can work through the math yourself, but in one year the total energy from such an installation will be just enough, at 100% efficiency, to produce three and one-half kilograms of the super-heavy element needed to make the *Impervium* alloy.[8] And, of course, at the power levels available in the '60s and '70s, a lot of those old-fashioned reactors were needed to produce the necessary amounts of *Impervium* to shield our missiles. Those reactors were built after much controversy, as you well-know, and a portion of the energy

[7] When I wrote this story in the late 1970s, a kilowatt-hour of electrical energy cost about 4 cents. Today (2025) it's about 16 cents. Thus, the electrical energy cost today of making a kilogram of the super-heavy element for the *Impervium* alloy would be about four billion dollars. By comparison, the cost of the entire WW2 Manhattan Project was about two billion (in 1940s dollars). The US does, in fact, routinely spend money at these levels on weapons technology. For example, the cost of a *single* B2 stealth bomber (the plane the US used to bomb Iranian nuclear weapons sites in 2025) is more than a billion dollars.

[8] Imagine you are one of the students in Professor Pixley's class. Can you confirm his numbers? (You can see how to do that in the box at the very end of this chapter.)

from each was diverted from the national power grid. The cover for all that was, quite simply, the so-called *energy crisis*! And that should also explain the 'failure' of Congress, in face of what was purported to be imminent doom, to do anything about it. People just wrote *that* off as the typical 'response' of Congress to most problems!"

The student who had started Pixley on his story had been squirming in her seat for several minutes, Finally, she could stand it no longer. "Look, professor, I don't mean to be disrespectful, but I just can't believe any of this. How could *everybody* be fooled? How could it be hidden from the government regulatory agencies that look at the records of the utilities? Wouldn't *someone* have revealed it all to the public eventually?"

"My dear young lady," replied Pixley, "it was the *government* that made the decision to manufacture *Impervium.* And *of course* the regulatory agencies knew perfectly well what was going on—*they* were the ones in charge of running the whole operation! As to the secrecy involved, the CIA manages to keep its multi-billion-dollar budget secret, even from Congress. Bureaucracies' can be very good at that sort of thing. During World War II the Manhattan Project, involving billions of dollars and many tens of thousands of people, maintained absolute secrecy. One of the slogans of the early days of the 'energy crisis' was that it was the 'moral equivalent of war,' only people then didn't really understand what was meant by that. It was for real!"

"Okay, Professor, maybe you're not putting us on. But why is it still hushed up? You say the Russians and we both have *Impervium* missiles today, so what's the point in not telling people what happened?"

Pixley banged his pipe into the empty wastepaper basket near the door and began to ream out the stem with a pipe cleaner. "The reason is simple. *It's still happening!* Oh, not so much as in the early days of the 'crisis,' and as the fusion plants come on-line the truth will probably be revealed in the next few years. You see, like all the other superheavy elements, the *Impervium* additive is unstable, with a half-life of only a few years. We have to keep making the stuff, even with a fixed number of missiles, to replace the missile skins as they lose their resistance to directed energy beams. The Russians have the same problem, of course, and that's why no one over there is talking, either."

The class stared at Pixley as he repacked his pipe with fresh tobacco. The boy in the front row finally spoke. "Professor, how do you happen to know all this? And how can the government be absolutely sure that the professor who discovered *Impervium* will remain silent?"

Pixley looked pleased with himself as he replied. "Let me answer your first question by responding to your second one with a counter-question. What makes you think he *has* remained silent? The government, however, isn't worried about the professor talking because who in their right mind would believe such a crazy story? Do you? Of course you don't. But it *is* true. I certainly wish, however, that I could find a way to convince you. And maybe I can!"

From the inside pocket of his expensively tailored coat, Pixley took out his wallet and extracted a thick wad of hundred-dollar bills. Spreading them out in his hands and holding the fan of money up for the class to see, he then snapped the bills back into line and wrapped them tightly into a tube along their long axes. Setting the tube on fire with his heavy, solid gold lighter, Pixley lit his freshly packed pipe with the flaming greenbacks.

"Believe me now?"

The backstory of a ballistic missile defense (BMD), in the form of space-based energy beam weapons, told by Professor Pixley to his students, is *not* strictly science fiction. The idea of such a system has been around in military weapons thinking for decades, going back to President Ronald Reagan's fascination with what became known then as "Star Wars." More recently, President Donald Trump has expressed interest in the idea of up-scaling the Israeli "Iron Dome" BMD from city-defense to a national system (called the "Golden Dome") to protect the entire United States. The year before "What *Really* Caused the Energy Crisis" appeared in *Analog*, I wrote a serious, tutorial-style paper on such systems: "The Laser BMD and Other Radiant Energy Weapons: Some Thoughts," *IEEE Transactions on Aerospace and Electronic Systems*, March 1977, pp. 96–107. It was the writing of that paper that motivated the story.

For a national BMD system "to work" in the case of a nuclear attack, the system would have to perform *flawlessly* in its very first use (there will be no do-overs and/or second-thought fine-tuning in a *real* attack!). That is, not even *one* of a barrage of incoming missiles can be allowed to "leak through" the BMD. As a benchmark on the likelihood of the Golden Dome achieving that perfect level of performance, consider the June 2025 Israel-Iran missile exchange: in one massive attack of 200 Iranian missiles, 22 of them penetrated the *extensively tested* Israeli BMD and hit their targets, a leakage of over 10%. Those missiles had conventional high-explosive warheads and, while causing great damage, that damage was localized in area. If missiles are armed with megaton nuclear warheads (as would certainly be the case in an attack on the United States), however, a leakage of even just *one* such missile would be a catastrophe (read the last line of note 9 in Chap. 2 again).

Calculating Professor Pixley's Numbers

The energy of a one-kilogram mass ($m = 1$) is $E = c^2$ joules or, $(3 \times 10^8)^2 = 9 \times 10^{16}$ J. Electric utility companies commonly measure energy in kilowatt-hours, where one kilowatt-hour is $10^3 \times 3.6 \times 10^3 = 3.6 \times 10^6$ J (because there are 3600 s in an hour). Thus, the energy of a 1-kilogram mass, in kilowatt-hours is

$$\frac{9\times10^{16}}{3.6\times10^{6}} = 2.5\times10^{10} = 25\times10^{9} = 25\text{ billion kilowatt} - \text{hours.}$$

Using a 10,000 MW generator (that is, a generator able to produce electrical energy at the rate of $10^4 \times 10^6$ joules per second, which is 10^{10} joules per second), then in one year the total energy from that generator is $(365)(24)(3600)10^{10} = 3$ 1.5×10^{16} J. This is the energy of the mass of *Impervium* additive that the generator can produce in 1 year, a mass of $\frac{31.5\times10^{16}}{9\times10^{16}} = 3.5$ kg.

14

Super-Science

Pulp magazine stories in the early days of science fiction were littered with super-science gadgets, with (*of course*) time machines at the top of the list. Other such gadgets included radios that could receive signals from the *future*, matter transmitters which allowed you to send yourself to a distant location just like we send e-mails today, ray guns, speech translators with which you can immediately talk with alien beings from anywhere in the universe, faster-than-light hyperspace drives, stasis-field generators which created regions of space in which there is no passage of time, and … lots more.

In today's science fiction, super-science gadgets are still with us:[1] what, after all, *would* authors do if they no longer had a time machine immediately available in the ink of their pens? Indeed, it is great fun for an author to "invent" a new super-science gadget, one that has never been used before. I have done that three times. One, of course, is the *photon duct* of Chap. 5 on invisibility. But before that I "invented" the *hyperspatial mega-time warper*, which appeared in the March 1980 issue of *Analog Science Fiction Magazine* (in the magazine's "Probability Zero" column). That column, reserved for short, really outlandish (the column's name is a big hint on that score!) pieces that all readers would instantly recognize as being beyond even far-out fantasy, is strictly for fun. Titled "Monster's Meal," my outrageous super-science contribution to "Probability Zero" is in the following box.

A more serious super-science gadget (certainly a lot more serious than is the hyperspatial mega-time warper) appeared earlier in my story "The Language Clarifier," published in the May 1979 issue of *Omni Magazine*. I say that the

[1] Matter transmitters, for example, appear on the very first page of Larry Niven's novel *Ringworld* (as "transfer booths").

© The Author(s), under exclusive license to Springer Nature Switzerland AG 2026

P. J. Nahin, *The Art of the "What-If?"*, https://doi.org/10.1007/978-3-032-15618-1_14

gadget in that tale is "more serious" because, unlike the hyperspatial mega-time warper, I suspect a language clarifier *will* eventually be constructed. Perhaps not by the end of next week but, with the accelerating development of artificial intelligence, almost certainly by the middle of this century.

The physics professor in the story was named in honor of my doctoral dissertation advisor (Jack Sklansky), about whom I'll tell you a bit more in the next chapter. "The Language Clarifier" is one of the exceptions to the disclaimer at the front of the book that all the story characters in it are not based on real people.

"Monster's Meal"

The whole unfortunate business occurred when Hubert Undersquat, science fiction editor extraordinaire, rejected Williard Pumpersnout's latest submission. In itself, a not surprising event, as Undersquat had bounced all of Pumpersnout's previous creations.

But this time he added a rather sour note to the standard rejection. "The scientific basis for this story is so obviously incorrect that even a fool could see it!"

This so enraged Pumpersnout (who fancied himself a rather brilliant inventor in real life), that he shortly thereafter appeared in Undersquat's office, bearing with him his latest gadget. "Call me a fool, will you? Well, take a look at *this*, Undersquat, and see if a *fool* could have conceived it!" And with that, Pumpersnout energized the Hyperspatial Mega-Time Warper. Within nanoseconds, both men materialized in a parallel Universe, populated mostly by telepathic monsters of voracious appetite and a pathetic weakness for a good after-dinner cigar.

Indeed, one of these slimy horrors was but mere centimeters from them as they popped into view. Reading their minds in a flash, the monster quickly engulfed Undersquat and ate him, all the while emitting a vast assortment of disgusting, slurpy-wet squeals of pleasure. Fortunately, none of these horrid noises can, in good taste, be reproduced here.

Turning to leave, now that its nauseating meal was done, the monster lit a cigar and left Pumpersnout untouched. "Wait, wait," cried Pumpersnout, "why not me, too? Why have you spared *me*?" Under the circumstances, a strange question to ask, you ask? Well, maybe Undersquat *had* been right about Pumpersnout's IQ.

In any case, the creature turned slowly back, blew a smoke ring in Pumpersnout's face and, with a spicy-flavored burp, replied, "You are a writer. And your companion was an editor. Correct?"

"Yes, that's true. But what has *that* to do with it?"

"Are not editors readers of manuscripts?"

"Yes, yes!" shouted Pumpersnout, "but so what?"

"Ah, you *must* be a fool not to see it. It is so simple. Readers digest. Writers cramp."

"The Language Clarifier"

The idea for the invention came during the divorce. He knew he was going to be screwed, but with the legal mumbo-jumbo of the separation agreements, he couldn't figure out *how* he was going to be screwed. Janet's damn fathead lawyer had drawn them up—he'd even given the go-ahead for that, as he hadn't planned to contest her. After all, he *had* been caught in a rather blatant, clear-cut position of adultery. At the time, he had thought the wild-passioned honey-blond had been worth it, but now he was beginning to have doubts.

He had a doctorate in semantics, and was the author of two scholarly tomes on the meaning and structure of words, bur Professor Willard Watson still couldn't understand what in hell was going on. Did he or didn't he get to keep the car? How about the house, the savings account, the cat and the dog, the antique hutch, the silver, the ski equipment, the home library, the television sets, and all the rest of the earthly possessions collected over 25 years of marriage? And what about alimony? Asking Janet's fathead lawyer led merely to the receipt of additional incomprehensible letters, notices, and other horrible documents. Just what the heck *did it mean* to receive a letter saying:

> *Notice is hereby granted to Willard Watson, the first party of aggravation with respect to the aggravated second party, Janet Watson, of an action for divorce, in the County of Orange of the State of California. Actions involved include but may not be fully delimited by their listing here: the exposure of the second party to loathsome disease by the first party due to participation in perverted crimes against the order of nature; public embarrassment of the second party due to the wanton, unrestrained, lascivious behavior of the first party; other acts committed by the first party of various horrid natures, to be specified at a later time, as needed, to describe the untenable position of the second party with respect to the first party. The second party maintains total freedom in the question of complicity of action, and, except in those cases where litigation proves contrariwise, sues for all common hereditaments, past, present, or future, to revert to the second party, except for the sole ownership of items, things, or other states of being in possession of the first party, prior to the initial date of marriage between the first party and the second party, except for such entities excluded by prior agreements in force at that time, or at other times not mentioned in this action.*

Professor Watson was somewhat perplexed by all this. So he hired his own fathead lawyer. What Professor Watson ended up with then was *twice* as much paper that he couldn't understand. Willard learned the truth of an old New England saying: *A man between two lawyers is like a fish between two cats.* So, he fired his fathead lawyer. And he stayed up for three straight nights, mulling over his desperate situation until the idea for the invention came to him. He quickly made an appointment to see his old friend at the college, Professor Sam Sklansky of the Physics Department.[2]

It was a cold, windy, and rainy day in early October as Willard ran from the parking lot to Sklansky's office. His shoes soon filled with water, and he squished his way up the steps into the Physics Building. Even Nature was dumping on him now. Sklansky's door was open, and he walked in, dripping, sloppy wet, with water slushing out his hat brim onto the floor. He stood there, looking like a lone, forlorn weed in the middle of a growing pool of water. Sklansky, a brilliant, very direct sort of fellow, looked quizzically back.

"So, what's the problem, Willard? And, by the way, umbrellas, raincoats, and boots *have* been invented. You some kind of health nut, running around in the rain like nature boy?"

"Look, Sam, I'm desperate, and I've had a lot of things on my mind besides the weather. I need your help, and I need it fast. Janet's going to rake my behind over the coals, but good, if I don't get someone to tell me what the divorce settlement she's serving on me *means*!"

"Willard, you want to see Professor Shyster over in the Law School. I deal in physical facts, mathematical validity, *in cosmic truth*, not the mental hash-mish-mash of lawyers."

"No, Sam, another fathead lawyer isn't what I need. I need *you*. I want you to tell me if my idea is possible,"

So, good friend that he was, Sam listened. At first he laughed hysterically, then he wrote a few Boolean equations and, seeing a little hope, he wrote some more. Then he became quietly excited and, finally, as Willard wrapped up his arguments, Sam became hysterical again. But this time it was with excitement. It *could* be done! The two old friends shook hands and agreed to begin construction that very weekend. Willard would provide a description of the necessary syntactical transformations, along with a complete table look-up dictionary of all the required synonyms, antonyms, and transitive verbs with irregular conjugations. Sam would provide the electronic expertise, produce the wiring schematics, order the parts, and do all the soldering.

It was just 2 weeks later that they stood in Sam's laboratory, looking at their gleaming creation. A cubical box, precisely 119 cm on an edge, it had a smooth, featureless appearance, with the double exceptions of 2 horizontal

[2] The inspiration for Sam was, as I mentioned earlier, my doctoral dissertation advisor, Jack Sklansky, who was an *electrical engineer*, not a physicist. Not wishing to have both Jack and Sam (with the same last name) in the same department, Jack's inspiration morphed into a physicist. Years after the story appeared, I asked Jack about that, and he said he was honored to be the inspiration for the brilliant Sam, no matter what department he was in!

slots. One was marked **INPUT** and the other **OUTPUT**. It was ready for testing.

"Okay, Sam, you designed it, you can have the honor of the first test."

"No, it was your idea, so you go ahead."

"Please, Sam, I insist."

"Well—all right, I *do* just happen to have a test problem ready." So saying, Sam walked over to his desk, rolled a fresh piece of heavy white bond paper into his typewriter,[3] and quickly snapped out in bold pica letters:

> *Liquid precipitation fell from the heights, followed by the spherical solid version, with the process terminated by the reverse transport in the gaseous state.*

Sam took the sheet over to the machine and, with an expression that was a mixture of glee and apprehension, held it up to the **INPUT** slot.

"Ready, Willard?"

At the nod of his friend's head, Sam pushed the paper in. After only a few seconds, another piece of paper shot from the **OUTPUT** slot. Both men grabbed it in midair, and together read:

> *First it rained, then it hailed, and finally the water evaporated.*

"Well, I'll be damned!" they exclaimed in unison. The Language Clarifier worked.

"Hey, hey, hey, Sam it looks good, it looks good!" Willard began to paw through his briefcase, looking for his divorce papers. "*Now* I'll find out what that scheming wife of mine is up to!"

"Wait, Willard," said Sam, as he placed a restraining hand on his friend's shoulder. "Let's not be hasty. We should really test it some more. Look here, I have a copy of today's campus newspaper carrying an interview with the Undergraduate Dean. Listen to this, will you, the perfect test!" He read aloud:

> *Even in institutions like our college, which may be expected to have rather homogeneous populations, one encounters a tremendous diversity in the family subcultures that students come from, in addition to the idiosyncratic mix of assets and liabilities that characterize them.*

"Wow, Sam—do we dare put *that* into it? It could blow the circuits!"

---

[3] If written today, of course, Sam would have a word processor and a laser printer in his office, and almost surely not a typewriter!

"Might as well find out if the Language Clarifier *really* works, Willard." Sam soon had the Dean's words typed in clear, crisp, sharp letters. He shoved them into the **INPUT** slot, and the machine responded in seconds with:

No two students are alike.

"Son of a gun, Sam, look at that! The translation actually makes sense. Try something else on it."

"Okay, Willard. Take a look at this—another quote from the Dean":

We thus encounter students whose educational aims are crustal clear, as well as others whose purposes have all the clarity of an amorphous mist emanating from a thick cloud of existential miasma.

Quickly they typed this out and inserted it into their machine, and they were soon in possession of the machine's response:

Some students know what they want, and others don't.

"That's enough for me, Sam—it works! Now, where the heck *are* those damn lawyer's papers!"

The rest is history. Willard finally found out what the divorce was going to cost him. He still got screwed, of course, but with the Language Clarifier deciphering the papers from Janet's fathead lawyer, he knew precisely *how* he was being screwed. Actually, Willard was really unconcerned, as he and Sam expected to make a bundle selling their machine to business, higher education, and government. Their need for extreme clarification was well-established. Let Janet have *everything*—secretly, Willard was *happy* to be rid of the damn cat and dog. He would recoup it all, and more, with royalties from the Clarifier. Willard let Sam handle the business end of the Language Clarifier, and it was with some greedy anticipation that he dropped in on him after the divorce was settled. Willard was flat broke.

"Okay, Sam, give me the news. How are we doing with the Clarifier?"

Sam opened his desk drawer, pulled out a piece of paper and handed it across to Willard. It was a cashier's check for 50,000 dollars. "There you are, Willard, your share of the proceeds from our first three sales. And more to come!"

"Hot damn, Sam, I knew it! Who bought the first three machines—businessmen dealing with government regulatory agencies?"

Sam grinned at Willard. "Professor Shyster, over at the Law School, bought all three."

"Of course," exclaimed Willard, slapping his forehead with a hand. "Lawyers *would* be the prime users of the Clarifier, wouldn't they? Why, with all the ritual chants they produce, they'll be in the market for Clarifiers for the next fifty years. What's old Shyster going to do with them, anyway?"

"Actually, Willard, you've got it backward. Shyster is writing a law book, and he's found that his early drafts weren't really up to par as far as the publisher is concerned. Not scholarly sounding enough, or something like that. So, the Clarifier is just what he needed."

"I don't get it, Sam," said Willard, with a puzzled look on his face. "If Shyster's book isn't impressively complex enough, how's the Clarifier going to help?"

Sam leaned back in his chair with a pleased look on his face. "Willard, my boy, there's an old rule of thumb in physics that says if a process works in one direction, it will almost always be true that it can go the other way, too."

Then Willard understood. "You don't mean, you couldn't *possibly* mean—".

"Yep, that's right. I just moved a couple of wires around, and now old Shyster just stuffs his clearly written book draft into the **OUTPUT** slot, and the most incomprehensible muddle you could possibly imagine emerges from the **INPUT** slot. Should be a legal best-seller. Shyster thinks the Harvard Law Dean will soon be calling with the offer of a tenured appointment."

Willard was stunned. The irony of it was mind-boggling. As he stared at Sam, his friend chuckled. "Look at it this way, Willard, how many of the lawyers who read it will *really* know, or even give a damn, whether they understand it or not?" Before Willard could respond, Sam's secretary put her head into the office.

"Excuse me, Professor Sklansky, but this large envelope, from Washington, just came for you registered, special delivery. It looks important, so I thought I should give it to you right away."

"Yes, good, thank you, Susan."

As the pretty young lady left, Willard found himself admiring the perfection of her departing form. "Careful, Willard," cautioned Sklansky, the always observant physicist. "As I recall, it was a blonde who did you in last time, and besides, she's the best damned secretary I've ever had. So stay away from her!"

"Ah, I suppose you're right, Sam, but she *is* a nifty-looking gal."

"Hmmmph," grunted Sam, who had been reading the just delivered mail. A sight frown was forming on his mouth. "Listen to this, Willard, it's from the Chief Legal Officer of Defense Research and Engineering in the Pentagon. Remember, I wrote to them about the Language Clarifier—pointed out how they could use it to decipher the thousands of proposals they get from industrial contractors every year?" He read:

Replying to your communication of 28 October, we have, after analysis of the broad ramifications of and pertaining to, in all its present and future forms, the Language Clarifier, found it to present a less than superior hold on the financial, economic, reputational, and any other forms of gain, physical or otherwise, of its inventors, in view of the the willingness of said inventors to receive and accept a yearly stipend, in perpetuity or for life, whichever terminates first, of one million dollars; they shall also accept the impact and import of the Military Secrecy Act of 1947, Title 12, Section 19.321 (see attached forms). Return of this document, with said inventors signatures, will constitute a mutually satisfactory agreement. Otherwise, not, with all applicable consequences to follow (see attached forms).

Sam put the letter down on his desk and drummed his fingers on the hard wooden surface.

"Well, Willard," said Sam, "what do you make of *that*?" Sam idly flipped through the 53 single-spaced onionskin pages of the 1947 Military Secrecy Act. "Frankly, Willard" he went on, "it sounds to me like the bastards are afraid to have the Clarifier around! You know, if the military boys can use it to blow away the industrial proposal-writer's crap, I suppose industry could use it to dig through all the government's crud, too. Why, *both* sides would have to make sense. Imagine that! And just think of the heart attacks in Congress when all the blowhards on Capitol Hill realized their speeches would be run through the Clarifier!"

"Hell, Sam, how the heck should *I* know? Look, let's run this letter and the Secrecy Act through the Language Clarifier—you still have our prototype unit in your lab, right?"

"Right, Willard. Let's go!":

A few minutes later the **INPUT** slot gobbled up the Pentagon letter. Then the 53 pages of the Military Secrecy Act of 1947 followed. A full **43 s ticked by as the Clarifier mulled over its latest task. Deep in its bowels a few transistors grew hot, an amplifier oscillated violently with feedback but recovered before vaporizing, and a mechanical gear-train drive almost ground off a tooth or two. But, finally, the Clarifier finished. It ejected its response:

Sign the agreement, forget you ever heard of the Language Clarifier, and you get a megabuck a year for life. Don't sign the agreement, and they toss you in the slammer (with one 60-second cold-water, low pressure shower every 10 days) and throw away the key.

Sam lives in Hawaii now, retired from teaching, and is writing a book on the physics of hanging ten. Willard also quit teaching, married Susan, and it would be indelicate to discuss what *they* are doing. Once a year they meet in San Francisco, split the million bucks, have a few drinks at Fisherman's Wharf, and ride the cable car.

Oh, yes, Sam was right. Old Shyster's book *was* a best-seller, and he's now being hotly pursued by both Harvard *and* Yale Law, thus proving you don't have to be smart to get paid a million bucks a year for forgetting what you know and doing nothing.

Quite often, merely being a fathead lawyer is sufficient.

The original title of "The Language Clarifier" was "The Anti-Amphibological Machine," using a word (guess which one!) that I found particularly fascinating. The then fiction editor at *Omni*—the late Ben Bova (1932–2020)—liked the story enough to buy it, but he wasn't enthusiastic about the title. He called me to let me know this, and I reluctantly agreed. There must have been something in my voice that told Ben of my disappointment, however, and he tried to take a bit of the sting out of the title change with his editorial lead-in when the story appeared in print:

> *Eschewing rhetorical obfuscation, he fabricated the perfect anti-amphibological machine!*

All these years later, I have come to see that Ben's decision was right.[4] But I *still* really like that word!

Finally, I do hope that all lawyers reading this story have not been too terribly offended. My reference to "fathead" lawyers was simply a story device. My own brother-in-law, in fact, is a retired criminal defense attorney in California, and he has assured me that I am still welcomed to visit (as long as I bring my own food and don't stay too long).

[4] Ben was a great editor: he received, *six times*, the *Hugo* award for the best professional editor, given each year by the Science Fiction Writers of America (the professional society of science fiction writers). In addition, he was a very talented writer himself. In a 2004 review of recent novels by Ben, *The New York Times* called him "the last of the great pulp writers."

15

Artificial Intelligence

In 1963, I graduated from Caltech with a Master's degree in electrical engineering. I was recently married, we had a baby on the way, and I was dead broke. Unleashing the massive brain-power expected of the typical Caltech graduate, I quickly concluded that I needed to get a job. Another big hint was that every time I turned the car radio on, it seemed to be playing that great 1957–1958 doo-wop song (that I remembered from my senior year in high school) by The Silhouettes: "Get a Job."

So, when the Caltech EE faculty asked everyone who was getting a Master's degree that year to take the preliminary oral exam required to go on for a PhD, I declined and gave my financial straits as an explanation. The EE faculty replied with "Well, okay, but just take the exam anyway." Not wishing to get on the wrong side of the people who hadn't quite yet actually signed-off on my Master's, I took the exam—and, not being at all invested in the outcome, I almost certainly turned-in the worst performance the EE faculty saw that year (if not, in fact, the worst performance of all time).

The end-result was that Caltech was not particularly sad to see me go, and I got a job.

Five years later, I was sufficiently solvent[1] that I could afford to return to graduate school and, in 1972, I *finally* got my PhD and started my academic

[1] In 1968, I was working as a Member of the Technical Staff at the Hughes Aircraft Company in Southern California. One of the perks for Hughes MTS engineers was the opportunity to apply for a Howard Hughes Staff Doctoral Fellowship, to use in a graduate program in engineering, physics, or mathematics. There were lots of great schools to choose from in the area, including the University of California campuses in Los Angeles (UCLA) and Irvine (UCI), the University of Southern California (USC), and, of course, Caltech (but I was gun-shy of Tech, as surely, I thought, they would remember my 1963 oral exam; how could they have forgotten it!). I picked UCI.

© The Author(s), under exclusive license to Springer Nature Switzerland AG 2026

P. J. Nahin, *The Art of the "What-If?"*, https://doi.org/10.1007/978-3-032-15618-1_15

career at Harvey Mudd College. But it was a close call. When, in 1968, I first approached the EE faculty at the University of California at Irvine, I was very lucky to be interviewed by the late Jack Sklansky (1928–2018), whose work on digital circuitry (done while employed in *his* pre-academic career at the RCA Labs in Princeton, NJ) that I had become familiar with during my employment at Hughes. Jack was a kind person, and sympathetic to my situation, but I have never forgotten his words: "You are 27 now, and will be 28 when you start. If you were just one year older, I wouldn't take you on. But, okay, let's see how it works out." Well, it *did* work out okay, but it was a *very close* call, and my life would be very different today if Jack had made a different decision.

My work with Jack (in medical image pattern recognition) was demanding (I still remember standing at the blackboard in his office, explaining what I had done in the past week, and his relentless requirement for a formal proof for *everything*). But it was also a very satisfying time. I did, however, hear some real horror stories, from my fellow PhD candidates, about *their* doctoral experiences with other faculty. Years later, the memories of those stories came together as inspiration for "The Box," which appeared in the January 1980 issue of *Omni*. I can only hope that the experience of getting a doctoral degree in electrical engineering, as described in the story, isn't a common one.

"The Box"

Ambrose P. Talaylee was bad news. At least that's what everybody in the electrical engineering department of East Pascatawa University thought. Why, the man was a disheveled monster, with unwashed hair awry, filthy jeans, straggly beard strung with fragments of yesterday's lunch, and his argumentative manner. He showed no reverential fear in the presence of his betters, what with his leaping at every opportunity to argue with one professor or another. He told vulgar jokes and laughed too much. It didn't help that he was brilliant. And he smelled bad, too.

Ambrose had appeared with a master's degree, 3 months before, from a small college in Ohio. He'd passed the entrance exams at EPU, and so they'd had to let him in. But right from the start the faculty had the jitters about him. Scoring a perfect set of examination papers upsets the evaluation process, you see. Of course, all the other grad students hated his guts since he'd made them look like klutzes. The next highest man got a 62.

At first, they thought they could get rid of Ambrose by stalling him. As old Wingate, the chairman, had roared in a department meeting, "Let the guy take courses until hell freezes over. He'll get no doctoral degree without a research dissertation project! Anybody here want to be his advisor?" They had all chuckled at the silliness of such a thing.

But Ambrose zapped them by finding a long-forgotten rule in the Graduate School Catalog. All he needed were the signatures of seven assistant Deans on a petition and he could go ahead with doctoral research on his own. The petition, dirty and torn at the edges, and covered with hideous-looking stains, made the rounds. Each A.D. had signed it just to get the strange, smelly person holding it out of their office.

No one really knew what Ambrose was working on. The petition just read "Implementation of a Synthetic Neural Network." As that cocksure young assistant professor Simpson said in another department meeting, "Maybe he's making himself a new brain!" That brought the laughs, and Simpson glowed with joy at having pleased his betters.

For the next 18 months, Ambrose labored in the former double-sized closet (lit by a single, naked, flickering 60-W bulb) that they had assigned to him. Occasionally he'd be seen in the stockroom, filling old paper bags with integrated circuit chips, capacitors, wire, and assorted nuts and bolts. He actually seemed to *live* in the closet, as some said they had heard strange noises coming from it late at night. The area around the closet door was constantly littered with paper trash like greasy pizza boxes, napkins smeared with catsup, and candy wrappers covered with chocolate stains. The man was a disgrace!

No one ever figured out where his money came from. Unlike the other graduate students, Ambrose had no teaching appointment in the sophomore labs. The smell, you see …….. And then one day Chairman Wingate called a special department meeting. "I have here a petition from Ambrose P. Talaylee, for the setting of a date for his oral doctoral examination." He held with obvious disgust, between two fingers, a ragged and soiled piece of paper.

Associate professor James, the department wise guy, quipped "Christ, Wingate, it looks like something died on that petition. If you've got any cuts on your hand, you'd better get a tetanus booster shot."

Assistant professor Simpson got his two cents in, too. "You mean he really got something done? Why, he hasn't talked to anyone in the department for advice or anything. All he's done is swipe the stockroom blind and smell it up something terrible!"

The chairman dropped the petition into a folder and gratefully pushed it aside. "Yes, but by the by-laws of EPU we must honor the request. If his work is a bust, we can deny his degree and run his butt out of here. So, who wants to be on his committee? There must be six people, you know. Two of us and four from other departments."

In the end, it was decided that Wingate and Simpson would be the electrical engineering representatives, while Professors Lardo, Hogson, Sweeney, and Vreeble, from physics, biology, political science, and mathematics would be the others. On May 17, at 1:30 pm, they all gathered in Memorial Hall, except for Ambrose, who came 40 min later. When he did finally sweep into the room, the committee members nearest the door reeled in his wake. Wingate and Simpson, knowing better, had sat as far away from the door as possible but, even so, they still experienced momentary nausea.

No one knew what to make of it. Ambrose was pushing a cubical metal box mounted on casters to the front of the room. He spoke just two sentences and then sat down. "This is my research project, a synthetic neural network endowed with all of Man's knowledge and the rules of logic to allow it to synthesize new knowledge. You many address all your questions to the box."

The committee members sat stunned. A mockery was being made of the time-honored tradition of oral doctoral defense! Just what kind of nonsense was this? But cagey old Wingate knew better than to make a scene. Otherwise, Ambrose could file legal complaints, and then all hell would be on their necks if the press got hold of it. Better to humor this obviously deranged person, let him make a fool out of himself, and then have a unanimous vote to deny the degree.

Wingate stood and walked up to the box. About 2 feet on an edge, its outer shell was black, crinkly aluminum. A combination loudspeaker-microphone was visible on the front panel. "Tell me," Wingate began, "what are the limits on the defining integral of the Laplace transform?"

With barely a pause, the box answered in a bell-like voice, "Lower limit is zero, upper limit is infinity."

Wingate was unshaken by that, as it was just a routine question. Any sophomore would have had that stashed away in his or her miniscule brain. A more impressive test would be to ask a partially ambiguous question to see whether the box could make plausible interpretations, as a human might do. "Correct. Now, what is the color of an electron?"

A slightly longer pause. "Color is the physiological sensation arising from a correlation between radiation wavelength and retinal response. Planck's law gives the relation between energy and frequency. To compute the color of an electron, it is necessary to know its total energy. Please input the physical environment of the electron, and then its kinetic and potential energies may be computed. This will define the frequency and wavelength and, hence, the color."

Wingate stared at the box. "Umph! Well, I pass to one of the other members of the committee."

Simpson leaped to his feet, shouting at Ambrose. "What kind of stunt is this, anyway? Have you got a smart midget inside that thing, Talaylee?" He forgot himself and stalked over to Ambrose.

Ambrose laughed in Simpson's face and Simpson staggered backward as a blast of rancid fumes engulfed him. Only his youthful vigor kept him from fainting. "Just ask it *anything* you want," Ambrose challenged.

Recovering, Simpson snickered to himself and decided to end this farce. "All right, box, try this on for size. Is there life after death?"

Quick as a wink came the answer: "Yes."

This disconcerted Simpson, but not for long. "Prove it," he yelled.

Back came the box with a response showing it was indeed Ambrose's creation. "Drop dead, and see for yourself!"

Simpson was shocked. "Why, why ..." he sputtered, "such arrogance, such ungracious behavior, is just intolerable!"

Ambrose laughed again. "I didn't say it; the *box* did!" The truth of that left Simpson speechless, and he sat down in defeat. The rest of the committee was reeling, too, but they retained enough of the committee instinct to go automatically into a huddle. They reached a unanimous judgment quickly, and gave Wingate the pleasure of announcing it.

"Mr. Talaylee, we have concluded that this box of yours is a complete fake. Its premise is absurd, and its answers are too rapid and lacking in sufficient scholarship. We have no wish to embarrass you further. So, we will not insist you show us the interior, where we are sure there is a radio transceiver linked to an accomplice elsewhere on campus. Therefore, it is our decision that, as regards your final oral doctoral exam, you fail! Good day, sir." With that, the committee marched out of the room in a column, with Wingate in the lead, like baby ducks following their mother.

Ambrose stood glaring after them, yelling "You bastards! I'll sue you. I'll sue!" Now that kind of response impressed the committee, even though it did lack a certain polish.

That night there was a secret meeting of the committee. At first, the professors thought they could still dump on Ambrose by awarding the doctorate to the *box* (after all, it, not Ambrose had answered all the questions). No, too risky. Instead, they threatened Ambrose with being blackballed if he persisted in his threat. "Look here, Talaylee," Wingate counseled (from a distance, of course), "if you'll just quietly leave the university, we'll call it quits. We'll even write you a good recommendation, one good enough to get you a teaching job at some reasonably decent high school. Get tough with us, though, and we'll see you never work in academia *anywhere*! Now, what do you say?"

"You stinking bastards! You're scared, aren't you? Well, you should be. I'm suing!"

For their defense, the committee hired Sylvester Shylock Shyster, a highly respected criminal lawyer. Ambrose elected to be his own lawyer. This is, of course, most unusual, but the judge relented after listening to Ambrose's arguments in closed chambers. Once the judge had caught the drift of Ambrose's arguments (to say nothing of catching the drift of Ambrose himself), he quickly agreed, and just as quickly he rushed from the room.

Since all the physical evidence was circumstantial, Shyster was confident of victory. But then the prosecution called its surprise witness—the box! Yes, the box had been in the laboratory/closet with Ambrose when the committee made its threats. The box had heard it all, but, after all, which of the professors cared what they said in front of a mere box?

Naturally, S. S. Shyster stood in court to protest.

"Your Honor, this is most unusual. This box is only a conglomeration of inanimate wires, circuit chips, and solder. It has no soul, no moral code of behavior, no knowledge of right from wrong. Its metabolism is that of amperes and volts, not that of a human's beating heart and the pulse of warm blood. It has no sensitivity, no ability to give voice to the feelings of humanity. Therefore, I move the testimony of this box not be allowed."

"Do you feel the machine is capable of lying?" the judge asked.

"No, your Honor. It has been shown to be a machine of computational infallibility. But the intricate complexities, the subtle nuances, of this incredibly convoluted case are just too involved for its mechanistic principles to grasp. It cannot, to make an analogy, appreciate the infinite variations that separate the pure white of innocence from the deep black of guilt."

"So," said the judge, "your objection really comes down to a feeling that the box may not be able to express itself in a sufficiently human manner. Well, let me ask the crucial questions, and we'll see if you're correct."

Turning to the cubical box (positioned comfortably on a purple velvet cushion in the witness chair), the judge asked "Tell me, box, did the committee, in fact, threaten to do damage to the reputation of your creator, one Ambrose P. Talaylee? And further, did the committee deny him the royal anointment of the doctorate?"

"Does a bear crap in the woods, your Honor?" the box answered. "Is the Pope Catholic? Is a hog's fanny pork?"

"I object," yelled S. S. Shyster.

"We confess," cried the committee in unison.

"Guilty as charged," the judge roared.

"Damn right," chuckled the box.

"Double damn right!" snorted Ambrose, who in his ecstasy of victory failed to notice that said snort caused all those nearby to turn green. But he wouldn't have cared even if he had noticed.

Thus did the committee learn, as its members were led away in shame and humiliation, of the error in treating an artificially intelligent box lightly.

It was later, as the committee sat glumly in jail while they waited for Shyster to come see them about making bail, that they learned things were even worse than they had suspected. Wingate had just read a note passed to him by the jailer. It was from his secretary. He hurled it to the floor as he screamed, "God damn it! Is there *no* justice left in the world?"

Professor Hogson from biology was stunned at the emotional outburst. "My heavens, Wingate, what's wrong now?"

Wingate stared at the floor, his jaw muscles working. "That bastard Talaylee! Not only does he get his PhD, but now he's putting us out of business. Do you know what that guy has gone and done? *Do you know?*" Wingate could see from the puzzled expressions on his colleagues' faces that they didn't know.

"I'll tell you what that creep has done! He's taken that miserable box of his, put a coin slot on it, and stuck it in the Student Union Building! He's got kids lined up getting their homework done for a quarter. It's charging fifty cents for an English essay, and two bucks for a master's thesis. A five spot gets you a doctoral dissertation!"

"My Lord, Wingate," gasped Hogson, who was near fainting, "academia as we know it will utterly collapse. What'll we do?"

"He's laid down terms. If we meet his terms, he'll stop." Wingate looked as if he were about to cry.

"Well, come on, man," croaked Sweeney of political science, "what are they? Whatever they are, we'll have to meet them!"

"He wants a faculty appointment at EPU, in my department," said Wingate, bursting into tears. "Oh, mother, I can't bear the thought!"

Professor Lardo of physics was the first to regain his composure. "Be tough, Wingate, be tough. So he gets an appointment. That doesn't mean he'll get tenure. Then, later, out he goes on his butt!"

Wingate slowly raised his head. "No, no, *no*, you idiot. You don't understand. Talaylee doesn't want the appointment for *himself*. He's already got a research job with the CIA. He can work alone for them breaking codes in a ventilated room, and so they don't care *how* bad he smells. It's that damned box of his. He wants the *box* to be given a faculty spot! Christ, a bloody damn computer as an assistant professor!" He lowered his head again, and began to weep quietly.

The professors sat motionless for several minutes, immobilized with despair, Then Vreeble, from mathematics, looked as if he'd just had an idea. "I say, Wingate, look at the *bright* side of it all." Wingate stared at him with a stunned expression on his face.

"Just what in hell is *bright*, Vreeble? It sure as heck isn't you!"

"Come, come, don't be bitter. See here, you'll be in excellent shape with the affirmative action, equal opportunity people. Gad, can you think of a better minority faculty hiring decision? Yours will be the first academic department in the entire country to hire a box! You'll be—".

Vreeble couldn't continue, for by now an enraged Wingate was attempting to throttle the unfortunate mathematics professor. Which is why, while the rest of the committee made bail and were released, Wingate had to stay another 6 months pending litigation of an attempted murder charge. But it all worked out okay, since it was Wingate's sabbatical year and he didn't have anything else to do anyway. It gave him plenty of time to figure out where things had started to go to hell and to ponder the evil of all prejudice, be it racial or nasal.

Everything worked out rather well for the box, too. It proved to have all kinds of good points; it was a great joke teller and was able to establish friendly relationships with students, and it didn't smell nearly as bad as Ambrose (unless the oil on its hinges got too hot). It was marvelous fun at department parties, and it never drank too much. One year later, when Wingate stepped down from the department chairmanship, the box was elected almost unanimously. Only Wingate and Simpson dissented. It's done just a great job since, too. There's even talk that when the Dean retires, then—well, maybe *that's* premature right now.

But still, who can tell? Who can tell?

16

Religion

Religion is a natural background for writers of both science fiction and fantasy. All three arenas require what is called a "suspension of disbelief," that is, a willingness to accept violation of a known physical law (indeed, maybe *more* than one violation). Writers have taken liberal advantage of that common feature, and the fictional literature abounds with science fiction and fantasy tales that have a religious connection.[1] By their very nature, such stories can have an enormously powerful impact on readers.

A short story that clearly demonstrates what I mean by that is "The Rescuer," by Arthur Porges, which appeared in the July 1962 issue of *Analog Science Fiction*. We read there of a time traveler who, after commandeering an experimental time machine, uses it to take a repeating rifle and 5000 rounds of explosive bullets twenty centuries into the past, to the day of Jesus' execution on the Cross. His intention? To shoot any Roman soldier who dares to even approach Jesus! Who, reading that, would deny immediately thinking "how wonderful doing that would be"? But on further thought, maybe things are not so simple. Porges develops the conflict faced by the time machine scientists who discover, to their dismay, the rogue time traveler's attempt to change the past. Yes, at first glance to save Jesus seems like a good idea, but what of the possibility that much more in history will also be changed?

There is time for the time machine scientists to intervene (in the story, it takes time to make a time trip)—but should they?[2] What if they don't? This

[1] For a book-length treatment of such tales, see my *Holy Sci-Fi!*, Springer 2014.

[2] A 1975 tale by Gary Kilworth, "Let's Go to Golgotha!", turns Porges' idea on its head. Rather than describing an attempt to stop the Crucifixion, in Kilworth's story the Crucifixion is a *tourist attraction*, a place to visit for cash-paying time travelers who are looking for entertainment! This powerful story is in Kilworth's 1984 short story collection *Songbirds of Pain*.

© The Author(s), under exclusive license to Springer Nature Switzerland AG 2026

P. J. Nahin, *The Art of the "What-If?"*, https://doi.org/10.1007/978-3-032-15618-1_16

concern about "changing the past" dates the 1962 "The Rescuer," of course, with the general belief today among physicists and philosophers being that demanding self-consistency in the physical laws of nature forbids such a thing from happening. *And yet*—such is the power of the religious nature of Porges' story that it is still a great read.

Even more shocking than the stories by Porges and Kilworth is the 1969 novella *Behold the Man* by Michael Moorcock. There we read of an emotionally disturbed time traveler who journeys back 2000 years in an attempt to find Jesus. Once in the past, however, he is stunned when he discovers that there is no one who matches the historical description of Jesus. So, the time traveler (who, before leaving on his trip in time, learned how to speak the ancient Aramaic language of the day, and had memorized every detail of the Biblical accounts of Jesus' ministry) *assumes the role of Jesus himself*—up to and including his execution on the Cross.[3] As he dies on the Cross the time traveler of course knows the brutal truth: he is no Son of God. He is simply a disturbed time traveler from the future who has acted out all the stories he had read in the Bible. Stories that will appear in the Bible only because *he* acted them out.

And so his last words before the grave, mumbled almost silently to himself and so unheard by his tormentors, are ones that surely stunned all of Moorcock's readers who are believers: "It's a lie … it's a lie … it's a lie."[4]

That is powerful stuff and I was, at first, convinced it would be very difficult for me to write something that would even approach the intensity of Moorcock's tale. My first attempt at a religious story was "Newton's Gift," which appeared in the January 1979 issue of *Omni*.[5] Like Moorcock, I have a time traveler journeying into the past—but not to find Jesus. Rather, my story is of a scientist who wishes to visit Isaac Newton, with the goal of presenting him with a modern, super-duper electronic hand-calculator (powered by a tiny nuclear battery that will last for years) to help the Great Man do his laborious arithmetic calculations. The twist in the story is that the time traveler is

[3] The story is relentlessly faithful to the demands imposed by self-consistent physics, demands that allow affecting the past but not changing the past.

[4] What happens to the time traveler's body after being taken down from the Cross is equally shocking, and yet consistent with the traditional Bible stories. I'll not spoil the shock for you here—read the story!

[5] At least I *think* it was my first attempt. Around the same time, I also wrote another "far-out" piece for *Analog*'s "Probability Zero" column (look back at Chap. 14 again), set in Heaven. The star of that tale is God, Himself, who is using perfectly fair dice (could there be *un*fair dice in Heaven?) to make all the random, quantum mechanical decisions continually being requested from all across the Universe (remember Einstein's famous comment on that?). It ended up being published in the *1981 Analog Yearbook 2*, which is probably impossible to find today. So, to end the book on an up-beat note, you'll find "Some Things Just *Have* to Be Done By Hand" at the very end of this chapter.

the cause of the well-known historical record of Newton eventually turning away from science and mathematics and toward religion. That redirecting of Newton's passion was the result of his misinterpreting the time traveler (bearing a "gift" with a display screen made of ominous-appearing, *red*-glowing LEDs) as being a messenger from Satan![6]

I liked "Newton's Gift,"[7] but had to admit its connection to religion is secondary to its presentation of paradoxes and causal time loops. So, I tried again with "A Father's Gift." It appeared in the August 1980 issue of *Omni*, and what I particularly like about it is the *lack* of a time traveler. It is a religious science fiction story with *no* reliance on a super-science gadget.

After reading "A Father's Gift," you might wonder if my inspiration for writing it came from watching the first Indiana Jones movie, the astonishingly imaginative *Raiders of the Lost Ark*? The answer is a resounding *no*: that exciting film (a true work of genius) was released in 1981, the year *after* my story appeared in *Omni*.[8] The Indy-like character in my tale does have some significant differences from Harrison Ford's Indy; my professor, for example, doesn't carry either a whip *or* a Webley Mk VI revolver (loaded with subsonic, man-stopping 0.455 cartridges, each the size of a large thumb). The most glaring difference of all, though (perhaps), is that I failed (alas!) to give my professor a fedora hat. *That* is the one, particular lapse I most sincerely regret.

Once you've read the story, I'll offer some final, personal, Indy-related comments.

"A Father's Gift"

I have found Christ.

No, no, don't say "Oh, one of *those* people!" Please, hear me out. I'm no zealous religious convert, no fanatic, not even a fallen politician seeking public absolution for misdeeds in office. I'm a hard-nosed computer archaeologist on the staff of a well-known university, specializing in the analysis of X-ray

---

[6] After co-inventing the calculus, after correctly formulating the laws of classical mechanics, after discovering the wave nature of light, after extending the binomial theorem to non-integer powers, after placing the inverse-square law of gravity on a firm theoretical foundation, after … (the list of Newton's scientific and mathematical discoveries is a *long* one), he then turned away from all that and embraced religious matters. In the end, while he wrote a lot on science and math, he wrote even more on religion. "Newton's Gift" offers an "explanation" for *why*?

[7] You can find it reprinted in my book *Number Crunching*, Princeton 2011.

[8] The actual inspiration for "A Father's Gift," and for setting the story in the Mummy Room of the Antiquities Department of the Egyptian Museum in Cairo, came after I read the 1973 book *X-Raying the Pharaohs* by James E. Harris and Kent R. Weeks.

axial tomographs of the mummies of Egyptian pharaohs. So when I say I've found Christ, I mean I've *found* Him.

And something else.

I've never been a devout man. That's led to some interesting discussions over the years with my brother Jack, who's an associate professor of ancient Middle Eastern languages at Georgetown University. A scholarly Jesuit, Jack had long ago kindled in me a fascination about the life of Jesus Christ *the teacher*. Can anyone doubt what a truly remarkable man He must have been? But I've never been able to accept the Church's dogma that He was the Son of God, the Savior here on Earth as the result of the Virgin Birth. And who, through the Crucifixion, suffered for the sins of all men. Up to now, that is, I haven't been able to believe. Now—well, let's just say I'm not so sure anymore.

We actually know so little about the life of Jesus, with what we do have coming only from the somewhat confusing, contradictory Four Gospels. We do, however, have a fairly good idea of the political times. It was the reign of the murderous King Herod (of whom Caesar Augustus once said he'd rather be a pig than a child in the House of Herod!), the Jews were oppressed by Rome, and the Children of God were eager for the coming of the Messiah long predicted in the Old Testament. The times were ready for a Savior, and Jesus Christ was the right man in the right time and place.

There is little dispute by scholars that Jesus was absolutely certain of His role. His life was no fraud, no shameful act of a charlatan. No actor would have suffered as He did. Some say He left us His image on the Shroud of Turin and nothing else. They say that He died and disappeared forever.

Or did He?

A number of respected Bible scholars (a minority, yes, but still a significant number) have questioned the traditional description of the death of Jesus, primarily because of their skepticism about the Resurrection. I must admit that has always been the stumbling-block, too, for my willingness to believe. Since I am an Egyptologist interested in funerary procedures, the death of Christ has fascinated me for years. It had always seemed to me that there just *had* to be an alternative explanation for what *really* had happened. And I was right.

It all started some years ago. While in Cairo at the Egyptian Museum, I was studying dental x-rays of their collection of royal mummies as a sabbatical research project. There I met a brilliant, intense man named Gamal el-Zam, now deceased. He was a professor of philosophy at the University of Alexandria, and was also on sabbatical leave for a year at the Museum's Antiquities Department. I became friendly with Professor el-Zam, and soon we were discussing our various research activities. Somehow the discussion got around one day to my interest in Christ, and of my conviction that His death was still a mystery no matter what the Bible may actually say on the matter. I recall he stared quizzically at me for a few moments, and I could see he was debating in his mind whether or not to pursue it. He must have sensed the depth of my interest, because he plunged on.

"So, my friend, you are a doubter, are you? Good! Possibly, then, you will find some papers I have to be curious reading. There is, I should tell you, great uncertainty about their veracity, as I believe they are actually part of a transcript of a previously lost part of the Apocrypha in St. Jerome's Vulgate. The Catholic Church, therefore, rejects them. But who is to say?—if you are as interested in pursuing the details of Christ's death as you seem to be, then maybe they will be of help. But I warn you—you may be getting into more than you bargain for!"

What he actually gave me wasn't the ancient manuscript itself, of course. The original has long since been lost, and what I received was a photocopy of the rediscovered part of the Vulgate. Included in that photocopy were the particular papers that Professor el-Zam thought would interest me. Before I tell you what was in them, it is best to start with the *Biblica*l account of what happened after the execution of Jesus.

After Christ was taken down off the Cross on the hill of Golgotha outside Jerusalem, His body was, according to the evangelists, taken by the wealthy Joseph of Arimathea to a nearby tomb cut in rock. After that, the body disappears, and the Gospel records become what unbelievers call *myth*, with its story of the Resurrection 3 days later, and of the ascent of Jesus into Heaven after 40 additional days. The Gospels are actually a bit shaky on this, because Luke also says Christ went up to Heaven on Easter Day, well *before* the 40 days were ended. That there are 40 days between the Resurrection and the Ascension has particularly fascinated me because Genesis, itself, mentions this as the usual time required for embalming. Could it be that the followers of Jesus spirited His body away from the tomb to prevent its defilement by the Romans, who might otherwise have buried it in a common criminal's grave? Could it be that an adherent embalmed the body of the Messiah, and then secretly buried it?

The long-lost papers given me by Professor el-Zam gave me the answers to my questions, for among them was a letter from Joseph of Arimathea to a man named Tertullian, apparently a close friend. First swearing Tertullian to secrecy, Joseph then describes the real fate of Jesus. The letter was in the ancient dialect of the common masses, Galilean Aramaic, which I could read only with great difficulty. Making an exact hand copy of the letter, but being quite careful to delete all references to Jesus by name, I sent it to my brother,

Jack, in America.[9] My wait for his response was agonizing, but it finally came 3 weeks later.

Greetings to my beloved, but unrepentant brother!

The strange text (where did you find it?) that you recently sent was most challenging. I enjoyed the mental exercise, but it has left me somewhat perplexed. When you return to the States, I want to have a long talk with you about it. But in answer to your request, here is my version of the original.

"And we bound his body in fresh linen and sealed his wounds. To secret its final fate, it was taken by night to a faithful follower, also a practitioner of the ancient art of preservation of the Egyptians. There it was purified, covered in soft lead sheets, wrapped in bandages, and sealed into a box of the Pharaoh. Transported overland to the Nile River, it was then sent by boat to the south, to the Temple of the Four Kings. There it was buried, safe at last from the Romans."

As I read Jack's words, words I had translated roughly myself but now was sure were right, I could barely contain my excitement. Jesus *had* been embalmed, His body mummified, and then secretly shipped to Egypt where it was then buried. But what was even more incredible was that I was also sure the coffin was no longer at its original burial site. I was sure I knew where it was, right now. Here, in the Egyptian Museum itself!

But I wasn't the only one to see Joseph's letter. Had anyone else reached the same conclusion I had? A conversation I had with Professor el-Zam, the day after getting Jack's translation, put my concern to rest. He was then about to return to Alexandria, and he inquired about my reaction to the papers he had given me.

"So, my friend, what do you think about those ancient documents, now that you've had an opportunity to examine them?" I answered carefully. Just how much did the professor really know?

"Gamal, they're fascinating. But one letter among them is the most fascinating, the letter from Joseph of Arimathea to Tertullian, concerning the events after the Crucifixion Have you read that one, Gamal?"

---

[9] *Omni*'s then fiction editor (Ben Bova—see note 4 of Chap. 14) was particularly keen on getting all easily checked details of a story right. It is okay, for example, to have a time machine in a science fiction story, but it is *not* okay to have a character in a story set in Kansas making his living working on a lobster boat. So, after Ben read "A Father's Gift," he wrote back to say "Was Aramaic a *written* language at that time and place? I don't think so." Well! That was a point I had completely overlooked, and my first thought was *I am in a lot of trouble (as far as selling this story goes)*! But then Ben's good friend, Isaac Asimov, saved me. Or, at least, his scholarly *Asimov's Guide to the New Testament* did, as it refers to the possibility that Matthew was originally written (around 70 A.D.) in Aramaic. After a bit more of looking into Ben's concern, I also learned that several non-Hebrew portions of the *Old Testament* (Daniel, Ezra, and Jeremiah) were written in a version of Aramaic. It was, therefore, at least *plausible* that Joseph's letter was written in Aramaic. All that was enough for Ben, and he took the story.

"Oh, yes, interesting to you no doubt, with your curious fixation on the death of Christ. But surely it cannot be authentic. After all, the expense of shipping such a coffin, so far, would have been enormous."

"Yes," I replied, "but Joseph was a wealthy man. He could have afforded it."

"I suppose, I suppose. But even if it is true, it must remain a conjecture. After two thousand years, buried anywhere in thousands of square kilometers of sand, the body of Jesus will have returned to the earth long ago. We'll never know. So, my friend, you now have another mystery to haunt you!"

I remained silent. The strange look that must have been upon my face no doubt was interpreted by the professor as disappointment. But it was something else—nearly uncontrolled thrill. Because I knew the professor was wrong! He was wrong because I knew the recent history of the Temple of the Four Kings, while Gamal was thinking only in terms of the ancient past. The Temple is more correctly called the Temple of Abu Simbel in Nubia, about 1100 km south of Cairo at the archaeological site of Gebel Adda. In 1960, when Nasser announced the plans for the Aswan High Dam, it was immediately recognized that the resulting floodwaters would drown Gebel Adda forever. A hurried salvage operation was thus started, in a desperate attempt to save what could be saved in the short time left. Among the artifacts recovered were more than 5000 human skeletons, *and several ancient coffins*, all of which were hurriedly cataloged and then shipped to Cairo. The skeletons have long since been extensively studied for bone and dental evolution over the centuries.

But not the retrieved mummies!

Considered to be just more Egyptian mummies among the many already scattered in the back of the Museum's second-floor gallery room, they had been ignored. It was standard policy of the Museum not to unwrap any mummy unless such a procedure was part of an ongoing scholarly study, and so the rescued mummies were simply subjected to a routine X-ray scan, after which the scans were filed and the mummy stashed away and forgotten. On my arrival at the Museum, to start my sabbatical, I had been allowed to go through the scan images of all the mummies, as part of my orientation. It was all pretty routine until one set of scans had briefly caught my attention—but the young woman in the records office had been unconcerned about my question.

"Miss," I recall saying, "this set of images is foggy. It almost looks like an under-exposure."

"Oh," she replied, glancing quickly at the images in my hands, "it looks more like the plates weren't aligned properly. The X-ray gun is placed on one side of the coffin, you know, and the film on the other—".

"Yes, yes, thank you. I am familiar with the technique." I returned the scan file and forgot about it, as her explanation seemed perfectly plausible at the time.

But as I sat silent before the professor, I now knew what the *real* explanation was. The film had been aligned properly, all right, but the mummy inside the coffin was wrapped in lead, just as the ancient letter by Joseph of Arimathea had said. The body of Jesus Christ was in that coffin—I was sure of it!—in a dusty storage room not more than a hundred meters from my office. How can I convey to you the excitement that charged my mind? The knowledge that the ultimate link between the modern world, and the long-ago world of a man who had changed the course of history, was in my possession. *There is no way.* I was like one overwhelmed by passion. I was to be the first in twenty centuries to gaze once more upon the features of the Messiah, Jesus of Nazareth.

The arrogance, the blasphemy of that desire, shames me now as I sit here in America recalling that incredible instant of understanding. But at that moment the lure was irresistible. It had to be done in secret, of course—how could I go to the Museum authorities as an outsider, an American on a temporary visa, and tell them they had Jesus Christ in their storage room? They would quite properly have had me locked up for observation. No, I had to do it alone. So, a week after receiving Jack's letter, under the pretense of working late, by midnight I was the sole inhabitant of the second-floor gallery. Armed with a crowbar and a heavy scissors, I made my way to the storage room. My heart was about to burst and my mind was reeling. I felt it was the greatest moment of my life.

- As he had done the year before with "The Language Clarifier" (see Chap. 14), Ben gave "A Father's Gift" a great editorial lead-in when the story appeared in print:

  It was the greatest discovery of the ages.
  All he had to do was open the coffin.

I had looked up the catalog number of the coffin with the foggy X-ray scans and, after about 20 min, I found it. It was in a dark corner of the room, covered with old packing crate materials and a layer of dust and dirt a few centimeters thick. It looked as if it hadn't been touched in decades. I soon had the coffin cleared away and began to pry the lid off with the crowbar. The lid was sealed solid with the ages, but my wild excitement gave me the strength of ten men.

I had the lid off!

Before me lay a mummy, wrapped in the usual bandages, apparently no different from any of the dozens of other mummies scattered about the Museum. I picked up the scissors. But then I looked more closely. In the dim light of the dusty corner where I stood, I had at first failed to notice that the bandages were charred, almost burned. And that they had a dull-gray glint to them, as if they had been sprayed with metallic paint. As I stared in wonderment at the strange sight, the scissors slipped unnoticed from my hand.

In awe, I stood frozen, unable to move. I felt I was in the presence of something that should be left untouched. But then I spotted an object lying free, nearly out of sight, at the foot of the ancient sarcophagus. I picked it up and, as I looked at what I had retrieved, a deep feeling of intrusion swept over me again even more intensely than before. I had found something Joseph had failed to mention in his letter. Pontius Pilate, as a mocking thrust at the priests who had condemned Jesus, had a wooden sign bearing the charge against Jesus nailed to the Cross. Joseph had removed Pilate's notice from the Cross, along with Jesus, and had placed both in the coffin, the sign to bear witness to the identity of the man it accompanied into eternity. Since it was written threefold, in Greek, Latin, and Hebrew, there is little doubt in my mind whose blood it was that had splattered across the ancient sign:

Jesus The Nazarene
King of the Jews

I slowly replaced the coffin lid, nearly overcome with emotion. Had I gone too far, had I pushed scientific curiosity beyond reason and into a region where it had no business intruding? I returned to my office, dazed, wrapped in agonizing uncertainty as I carried the bloody sign with me. But still, I couldn't let it be. I recall staring at the sign for hours, there in the gloomy quiet of my office. I *had* to learn the secrets of what I had found.

The blood! It was all that was left, the only physical remains I had of Christ's body—except for the mummy itself, which I hadn't had the courage to pry from its wrappings. So, once again I wrote to America, this time to ask a friend in the pathology laboratory of the medical school at my university, an expert in forensic medicine, to run a total blood analysis on a fragment of wood I happened to have. Of course, I told him nothing of its origin. I don't know what I expected to learn, but I couldn't help myself. I was so close to understanding the mystery that had haunted me for so long that I just couldn't leave it alone. And the blood was all that I had. I soon had my friend's report. Halfway through it I had to put it down in shock. It couldn't be! But if it was so, then what could God have intended by allowing His Son to be so cursed?

The lab analysis had found the presence of a large number of distinctly malformed lymphocytic cells, supporting a diagnosis of lymphosarcoma. Today we'd treat it with Amethopterin and 6-mercaptopurine, with the usual result being a total remission, one possibly lasting as long as 5–7 years. Two thousand years ago, however, it was a virtual death sentence, with a survival time of no more than 8 weeks. If Christ had not died on the Cross, He'd have been dead within 2 months anyway. Or so I thought as I read the analysis.

Such a death would have destroyed the perception of Him as the Messiah in the eyes of His followers. So, in that sense, the Crucifixion came just in time. But why the acute leukemia? Why a disease of imperfect man in the body of the Son of God? As I pondered those questions, the second half of the blood analysis contained a second surprise, one presenting a riddle that, in its own way, is as perplexing as is the first.

My friend's delicate chemical tests had also detected the presence of certain toxic blood reaction products; products produced *only* in response to the modern drug treatment for leukemia. It was as if He'd been on chemotherapy, and was beginning to suffer a relapse just before His execution! I sat stunned, numb with disbelief: it was so incredible. I pulled open my bottom desk drawer, the one I had started locking since hiding the sign in it. As I held the old wood in my hands, I began to doubt. Was the sign *really* old, or was it all just a fantastic hoax? Was Professor el-Zam merely making me the butt of an elaborate, cunning prank? I didn't think so, but …? I knew then what I *had* to do, what I *must* do, if I was to know another moment free from confusion.

I had to cut open the mummy!

The very next night found me again in the storage room. I was then almost in a fever and so had the coffin lid off in just a minute or two. I then immediately attacked the strangely glinty, burned bandages with my scissors. But what!—the ancient cloth fell apart at the first thrust of the blades, and what was revealed to me was so astounding that I dropped the scissors and staggered backward.

When I had begun to cut, the cloth had been in the form of a heavily wrapped figure. As the cloth separated, however, under the force of the blades, the fabric fell in on itself as if it contained nothing but empty space. And indeed there were no bones, *nothing at all*, to mark the resting place of a man. But then I saw that there *was* something there. A congealed, roughly spherical ball of metal. It had to be the lead mentioned in Joseph's letter, the lead that had been wrapped around Jesus. I stood paralyzed by both wonderment and surprise. There no longer was any doubt in my mind: this *was* the coffin of Jesus.

But where was He?

Slowly, the connection between the charred bandages and the melted lead sheets became clear in my bewildered mind. Something had caused the temperature of the covered body to reach at least the temperature of molten lead—328 °C. If the heating had occurred in just the right way, the surface tension of the liquid metal would have pulled it into a sphere. And a lead mist would have impregnated the bandages, which in turn would have fogged an X-ray scan.

I thought to myself: *it would have taken energy, wouldn't it—a* ***lot*** *of energy—for Him to return to His Father!*

The mystery that had bedeviled me for so long was finally resolved. The bloody sign, with its incredible tale, and the ancient coffin, had shown me the way. While dying on the Cross, Jesus had cried out in agony "My God, my God, why hast thou forsaken me?" If only He could have known then, in His moment of extreme anguish, that His Father had *not* abandoned Him, but rather had given Him the gift of life beyond His natural time on Earth.

As my thoughts dwell today on my discoveries in the Egyptian Museum, and as I think of the compassion of the Lord God Almighty for His Son, I feel comfort and warmth. I feel at real peace with myself for the first time in my life. I *didn't* do wrong in pursuing the riddle.

I *have* found Christ.

After filming the sixth Indiana Jones movie (*Dial of Destiny*, 2023), actor Harrison Ford declared that he was done playing Indy battling villains, all while in pursuit of some religious object. He was, he said, now simply too old for that sort of thing. That declaration struck gloom into the hearts of all true Indy-fans (mine included). That gloom only deepened with the decision by The Disney Company to use extensive computer-generated graphics to make any future Indiana Jones films. So, is that the future, a *software* Indiana Jones rendered by terabytes of computer code? Well, *perhaps*. Or, maybe not. I do think "A Father's Gift" leaves the door open for one more *plausible* adventure, with an *elderly* Harrison Ford as Indy.

But let's take Ford at his word that he is done with Indy, while we also realize that Indy is *not* the only archaeologist who pursues religious artifacts. They *all* do! The situation is similar to the fact that even though Humphrey Bogart was great as film-land's man-of-action, hard-boiled private detective Philip Marlowe in *The Big Sleep* (see note 7 in the Prelude), there was still lots of room in the detective genre for the elderly, intellectual, scholarly sleuth created by Agatha Christie—Hercule Poiret—as played by David Suchet on television for nearly 20 years. My professor in "A Father's Gift" is the Poirot version of Ford's Indy.

So, as the film I envision opens, we find the professor delivering a guest lecture at the Egyptian Museum in Cairo, on his fascination with the death of Jesus. Then, in scenes yet to be developed, he is caught-up in pursuing the secret of the long-lost coffin of Jesus—and what greater religious artifact than that could there be?

Well, no matter who plays our archaeologist, for me it is the *Egyptian Museum*, itself, that irresistibly beckons … and—Holy Cow!— could that be a Satanic demon, also in pursuit of the coffin, that I see lurking in the shadow cast by the nearby casket of Queen Henttowy of the Twenty-First Dynasty? Come back next week, kids, and find out!

Finally, to wrap this book, here is the "Probability Zero" tale from *Analog* that I mentioned in note 5, a tale that asserts there are things *even in Heaven* that could use some improvement.

"Some Things Just *Have* to Be Done By Hand"

The Most Important Entity rubbed His temples in fatigue. There was just so damned much crap to put up with nowadays. The personnel paperwork was nearly overwhelming, even for a being with omnipotent powers. And a work force faced with zero turnover had a first-class morale problem. The younger ones knew there was no hope for advancement by the once-usual routes of death, retirement, or resignation. None of those events ever happened—here. The telephone rang, and He answered in weary relief at the distraction.

"Yes?"

"Sorry to bother you, Sir, but the main computers have a backlog in the RANDOM QUEUE for ten to the 183rd power decisions. Can you please service those requests right now?"

"Damn, are those bloody scientists on Earth doing their quantum experiments again!? You'd think they'd understand the Uncertainty Principle after all these years. Well, what is it now, an electron beam through a diffraction grating, or is somebody trying to locate an atom with zero error?"

"Both, and more, Sir. Those guys are really getting busy down there. Why, just as we've been talking here, the RQ has picked up ten to the 179th power more requests!"

The main computers couldn't be allowed to overflow. Once, two or three thousand years ago (in Earth time), they had been unattended for several days (in His time), and the RQ had clogged up tight with ignored decision requests for determining the outcomes of random events. The resulting massive computer system crash had caused entire centuries (in Earth time) of strange, abnormal violations in His Laws of Natural Phenomena. It had been the time of magic on Earth, and the new wizards, sorcerers, and magicians had used it to their advantage in proclaiming themselves to be all-powerful. It couldn't be allowed to happen again!

"All right, all right, hold your feathers smooth. Hang on for a moment."

He put His caller on hold, and pulled open the desk drawer next to His perfect left foot. Inside was a pure diamond crystal box, containing two ruby cubes of ultimate clarity. The dots on the cube faces were precise circles of gold. The cubes were perfectly balanced, of course, as it was impossible for anything unfair to exist—here. Taking the cubes in His mighty hand, He established a mind-link with the input-output data lines to the main computers. Faster than imaginable (or even possible by ordinary laws, but for Him very little was impossible) the cubes tumbled in His quivering hand. The whole thing was over in just a few wing beats. He dropped the cubes, now so hot they glowed in the gamma-ray region of the spectrum, back into their crystal box, and shoved the drawer shut with a kick from His perfect left foot.

"Okay, the main computers cleaned up?"

"Yes Sir, the RANDOM QUEUE is empty!"

"Excellent—now please don't call again for at least another day. Meanwhile, you and your colleagues might busy yourselves with finding a way to speed up the automatic software random number generator. I find this business of hand generation to be increasingly inconvenient. Good-bye."

As He hung-up, He thought of what Albert Einstein, one of the better Earth scientists, had once said: "God doesn't play dice with the Cosmos."

"Humph," He grunted in disgust to Himself, "just what the Hell did he know about it?"

GPSR Compliance
The European Union's (EU) General Product Safety Regulation (GPSR) is a set of rules that requires consumer products to be safe and our obligations to ensure this.

If you have any concerns about our products, you can contact us on

ProductSafety@springernature.com

In case Publisher is established outside the EU, the EU authorized representative is:

Springer Nature Customer Service Center GmbH
Europaplatz 3
69115 Heidelberg, Germany

www.ingramcontent.com/pod-product-compliance
Lightning Source LLC
Chambersburg PA
CBHW060404040326
40615CB00024BA/875
* 9 7 8 3 0 3 2 1 5 6 1 7 4 *